The Financial Times Essential Guide to
Making Business Presentations

The Financial Times Essential Guide to Making Business Presentations

How to deliver a winning message

Phillip Khan-Panni

PEARSON

Harlow, England • London • New York • Boston • San Francisco • Toronto • Sydney
Auckland • Singapore • Hong Kong • Tokyo • Seoul • Taipei • New Delhi
Cape Town • São Paulo • Mexico City • Madrid • Amsterdam • Munich • Paris • Milan

PEARSON EDUCATION LIMITED

Edinburgh Gate
Harlow CM20 2JE
Tel: +44 (0)1279 623623
Fax: +44 (0)1279 431059
Website: www.pearson.com/uk

First published in Great Britain in 2012

Pearson Education is not responsible for the content of third-party internet sites.

ISBN: 978-0-273-75799-3

British Library Cataloguing-in-Publication Data
A catalogue record for this book is available from the British Library

Library of Congress Cataloging-in-Publication Data
Khan-Panni, Phillip.
  The Financial times essential guide to making business presentations : how to deliver
a winning message / Phillip Khan-Panni.
     p. cm.
  Includes index.
  ISBN 978-0-273-75799-3 (pbk.)
  1. Business presentations. I. Financial times (London, England) II. Title. III. Title:
Making business presentations.
  HF5718.22.K46 2012
  658.4'52--dc23
                          2011030717

10 9 8 7 6 5 4 3 2 1
15 14 13 12 11

Typeset in 8.75/12pt Stone serif by 30
Printed and bound in Great Britain by Ashford Colour Press Ltd, Gosport

This book is dedicated to my wife, Evelyn, who made it possible for me to write it swiftly when there were so many external pressures, and made the space for my creativity

Contents

About the author

Phillip Khan-Panni is CEO of PKP Communicators, providing training in communication skills, cross-culture and salesmanship in person and in print.

A professional speaker, he has won numerous public speaking titles, including several international championships, UK Champion in the World Championship of Public Speaking (organised by Toastmasters International) a record seven times, and is the UK's first and only World No. 2. He is a co-founder of the Professional Speaking Association, and was its Marketing Director for the first four years.

He is a past Branch Chair of the Chartered Institute of Marketing.

Phillip's business background is in speciality selling, advertising and direct marketing. He was Senior Copywriter at *Reader's Digest*, London, for eight years. On the London *Evening Standard*, he set up the Sales Promotion Department, and on the *Daily* and *Sunday Express* newspapers, he was the most successful Classified Advertising Manager in their history, tripling ad revenue in his first year.

Phillip was Managing Director and Creative Chief of PKP Communications Ltd, a Direct Marketing creative agency with blue chip clients, until a family tragedy caused him to close it down.

Since 1994 Phillip has been coaching senior business executives in public speaking, presentations and cross-culture, work that has taken him to 20 countries around the world.

He has published seven previous books on communication skills, and one of poetry. This is his eighth business book.

Introduction

This is not just another book on presentation skills. Far from it. This book gives you insights that could transform your business presentations and significantly improve your conversion rate. It has been constructed to act as a reference source for the present, and a coach for the future, providing checklists, tips and guides to help you maintain high standards.

In fact, it is about persuasive communication.

This *Essential Guide* is intended to give you an understanding of what it takes to *get results from business presentations*, and how to prepare and deliver your message with impact. It includes the relevant elements of oratory and the platform skills of a professional speaker. It will help you to grab and hold the attention of a business audience.

The focus of this book is on helping you to understand *how persuasion works*. Its premise is that every business presentation should be designed to make some change in the thinking, attitude or behaviour of the audience. It therefore follows the rules of salesmanship, and includes the techniques of powerful platform performers.

Let's be clear about the aims of this book. It is not about selling and it is not just for those who work in sales, but it *is about getting agreement*. It's about making a proposition and getting that proposition accepted and acted upon. It's about making your case in a way that gets results.

Presentations that worked

Starting with examples of successful presentations, *Making Business Presentations* examines why they worked, and contrasts them with presentations that fail, emphasising the costs of getting it wrong. The book provides some simple techniques, but explains the reasoning behind them, so that you can make lasting improvement and monitor your progress.

When you know what success looks like, and realise the contribution made by good presentation, you will never again consider a business presentation to be just a parade of PowerPoint slides. It will make a difference to your self-confidence when you stand before an audience.

In Chapter 2 there is a checklist of *30 essentials for successful presentations*. It will help you to understand why so many presentations fail . . . and what you need to do to avoid making the same mistakes and improve your own results.

You'll be guided to realising your own added value in the process, and the importance of your own point of view in the messages you put across.

One essential element that will make a huge difference to your business presentations is this: stop describing your product or service, and talk, instead, about how it benefits your customers.

What you will learn

You'll learn a simple technique for going from *blank page to first draft in 15 or 20 minutes*. It will get you started, it will help you get out of trouble when you have a tight deadline, it will provide a working framework for all your future presentations, and give you an insight into the value and purpose of structure.

You will gain an understanding of how people listen to presentations, so that you can connect better with them and avoid 'death by PowerPoint'.

Aids to memory

There are a number of aids to memory, such as the 4 Ps of preparation, which apply to presentations: *Principles, Purpose, Planning* and *Pre-qualifying*, as well as the *three negative elements* and *three positive elements* of presentations.

There is a whole chapter on persuasion, to help you make your case more effectively. Many business presentations concentrate on describing the offer or the nature of the business they represent. I show you how to follow the sequence that presses the right response buttons in your listeners.

The chapter on visual aids will guide you away from the slide abuse that ruins presentations, explaining how slides and other visual aids should be used to support but not supplant you, the presenter.

An extension of what you already know

This is a book you can read from cover to cover or dip into from time to time. Each chapter highlights the learning points, to enable you to refresh your memory as you go along.

I have assumed that you already know something about making business presentations, but have added some new thinking and assembled it all as a handy ready reckoner – with the appropriate explanations where necessary.

I have also included something on the oratorical devices that are used to such great effect by such brilliant speakers as Barack Obama. They are known and proven to work in pressing people's hot buttons, and can make an amazing difference to the way you connect with your audience.

The final chapter is about revision and setting up a process to monitor your future performances as a presenter. It provides a quick-reference summary of all the chapters, to enable you to locate any of the main topics covered. In addition, it provides access to the website where you can download a recorded summary of the book, chapter by chapter, and burn it onto a disc to play in your car and elsewhere.

part

1

Planning

Successful presentations

In this chapter

- Dramatic launch success through 'fun' presentation
- Powerful demonstration that created a worldwide phenomenon
- Engaging the emotions
- The magic of great oratory

Everyone in business makes presentations, often without realising they are doing so. A business presentation does not have to be a slideshow. In simple terms, it is whatever method you choose to illustrate, demonstrate or explain what you do, in order that the other person will 'buy' it.

It could be a speech without notes, a talk with slides, a demonstration, a stage act, a group performance. Whatever form it takes, it's a sales pitch, pure and simple. It says (or should say): 'This is what I can offer. Would you like some?'

There is enormous power in words and even more when they are combined with demonstrations that prove their intent. We can use them to make a sale, to get a job, to gain influence, to make our reputations, and even to change the course of history. Let's consider some remarkable achievements that came about because of the way in which the proposition was put.

Live test

In 1998 three Cambridge graduates working in the City of London had an idea they wanted to test at a London music festival. They bought £500 worth of fruit, whipped them into smoothies, and sold the new product over the festival weekend. Their testing vehicle was an ingenious one, which has since been much copied.

They asked customers to put their empty bottles in a 'Yes' bin, to indicate if they would buy it again. The three thought they would quit their jobs if the test proved successful.

They put their products in clear plastic bottles with paper labels, and decided to have some fun with them.

Printed on the bottom of the bottle itself were such comments as *'Open other end – it's easier'* and *'Stop staring at my bottom'*. On the back of the label (visible when the bottle was empty) was more fun copy. By the list of ingredients was the notice *'Separation may occur**'. At the bottom of the label was this: *'*But Mummy still loves Daddy.'*

Not only did they have a new product, they had *a new way of addressing customers*. It established the brand right away. Customers enjoyed the product, but they also enjoyed the way the product was being presented, and it encouraged them to enter into a relationship with this new outfit.

Quick success

The 'Yes' bin was filled so quickly that the three young men resigned their jobs the next day, and started a company which they called 'Innocent'. Today they sell two million smoothies a week to over 7,000 shops, and their annual turnover tops £100 million. The initial demonstration answered three questions:

1 Did people like the product?
2 Would they pay for it?
3 Would they buy it again?

Once again, the presentation was a demonstration of the product, wrapped in a research format. Interestingly enough, people seem more likely to try a product (providing the cost is nil or low) if it is represented as a 'trial'.

Having the 'Yes' bin there enabled customers to try the product and register their views immediately. The key element here was 'involvement' – something that plays a part in most successful presentations. The involvement was partly through the 'testing' process and partly through the witty labels, i.e. the presentation.

The power of demonstration

One of the most dramatic, breathtaking presentations took place on stage in 1994, and resulted, within a year, in sales and an international following measured in millions. Although you may not have considered it a business presentation, it proved the power of demonstration, and created a significant business that is still thriving. That was *Riverdance*.

It came about almost by chance. As winners of the 1993 Eurovision Song Contest, Ireland hosted the 1994 event, and the producer,

Moya Doherty, invited two American-born traditional Irish dancers, Michael Flatley and Jean Butler, to choreograph an Irish dance number to be performed during the interval, while the judges voted. The MC was Terry Wogan, who announced that we were in for something very special.

By the end of the tap dance act, the hairs on the backs of people's necks were standing on end and the live audience were on their feet, applauding wildly, eyes moist with emotion. They had never seen anything like it. Grinning with delight, Wogan murmured, 'Tell me you didn't like that!'

The measure of success

Bill Whelan's composition, *Riverdance*, outsold the winner of the Eurovision song contest that year, and in 1995 the act had been expanded as a full stage show, going on to make millionaires of the originators.

What made it so successful was a combination of several fundamental factors which have a bearing on business presentations. It was different, it was perfectly executed, it was launched at the right time, and in the right circumstances. That short performance during the interval was a presentation of the *Riverdance* phenomenon, a performance that demonstrated its audience appeal and its commercial possibilities. No claims about it were necessary.

In the world of advertising it has long been said that it is always better to demonstrate than to claim, and *Riverdance* proved the truth of that saying. The live audience and the millions watching on television said, implicitly, 'We'd like some more of that!' That was success.

Essential tip

▨ It's better to demonstrate than to claim.

Classified results

Often a presentation can be the start of a sales campaign. If it is done right, its impact can carry over into the new campaign, instead of just becoming the fading memory of an event. In my mid-20s I was given a short-term assignment on the *Daily Express*, to lift the classified ad sales out of the lethargy that had stunted its growth. It was in the bad old days when advertising salespeople ate lunch for a living and were little more than order takers.

Some of the salespeople had 'flip over' presentations in black folders, which seldom saw the light of day, as they thought it sufficient to inflict their personalities on their prospects. Classified advertising was a very unrefined art in those days. I redesigned the presentations, placed the City Golf Club 'out of bounds' and trained the sales team in more positive techniques, such as the *Five Questions to Answer Before Making the Call*. In three months I increased revenue by 33 per cent.

Three years later, ad sales on the *Daily Express* were again mired in complacency. The Managing Director, Jocelyn Stevens, decided to relaunch the *Express* and asked me to return and *double the classified ad revenue in a single year*.

Advertising agencies had already been invited to attend a presentation, for which some embarrassingly bad slides had been prepared. I filed them in the bin and created a new, lively presentation using twin projectors, music and two presenters – the Advertisement Director and myself.

Expecting a low-grade performance, Jocelyn Stevens was reluctant to attend the presentation, but at the finish he was left with his mouth open at what he saw and from the feedback from the invited agencies.

We set and broke more records than ever before or since, established a credible alternative to the market leader, *The Daily Telegraph*, and *TRIPLED the revenue in just ten months*.

Reasons why

What was it that enabled such success?

- We focused on the needs of the market: a credible alternative to the market leader.
- We understood our own product better.
- We supported our advertisers, providing our readers with incentives to respond to the ads we carried.
- We stopped selling 'space'. Instead we spoke of the numbers of the right kind of readers we could offer – people most likely to respond to the ads.
- We got results for advertisers.

It all began with one presentation, but the entire sales effort was a continuous presentation of what we could do for advertisers. Every salesperson was equipped with a well-reasoned case for being seen by media buyers. It was not a rigid script, but there was a planned sequence to ensure that the salesperson knew just why they were there and why the media buyer should see them.

It was, in effect, a presentation based on the concept of *a joint venture*: the advertiser needed readers of a certain kind, and we offered those readers plus the right environment in which they would see the ads and respond.

Not only did we increase revenue at a rate never matched before or since, but some of the salespeople I trained went on to become directors and senior managers in newspapers, radio and advertising agencies. They told me they applied the same thinking in the presentations they, in turn, devised and trained others to produce and deliver.

Essential tips

- Know how your offering benefits customers.

- Believe in it.

- Follow a disciplined approach.

- Treat a pitch as a joint venture.

Engaging the emotions

In September 1994, the British medical journal *The Lancet* caused a stir when it published an article urging doctors to apply the techniques of acting in their work. Drs Hillel Finestone and David Conter of the University of Western Ontario wrote that doctors should be trained as actors. They stated: *'If a physician does not possess the necessary skills to assess a patient's emotional needs and to display clear and effective responses to those needs, the job is not done.'*

It's right in medical practice and it's just as right in business. *When you engage the emotions of those with needs that you can satisfy, you get 'buy in'.*

When colour advertising came to newspapers it was much more expensive than black and white, so I knew we had to engage the emotions of the media buyers. Did you ever see the film *The Glenn Miller Story*, in which Glenn Miller, played by Jimmy Stewart, searched for an elusive 'sound' that he could call his own?

In the film, when he first performed with that sound, Miller (Jimmy Stewart) focused his gaze on the feet of the customers. When they started tapping he knew he was in business, with the distinctive and irresistible Glenn Miller sound.

That was the inspiration I needed, prompting me to build two elements into my presentation. The slides showed a succession of newspaper ads in black and white, and then in colour. Each ad was shown first in mono, then in colour, then again in mono, and colour once more. The audience could see for themselves how much better the colour versions were. They were not left to imagine the difference or the benefit.

One of the featured ads was for a red car. The contrast between the mono version and the one in colour was quite dramatic, and you could *feel* the rise and fall of emotional response in the room as the slides switched between the mono and colour versions. But while the ads were running on the screen, we played a recording of a catchy tune, *Popcorn*.

Within minutes, people were tapping their feet. They were engaged! When people are engaged, they become more responsive.

Essential tips

■ Engage the emotions.

■ Show the benefit.

Music can improve response to ads

Colour printing produced a surge in advertising revenue, not only for the papers where I worked, but across all national newspapers, and accelerated the development of the printing technology that made newspapers so much more attractive.

Some years later there was some research done at the University of Florida on the effect of music on emotional response to advertising. Jon D. Morris and Mary Anne Boone found that music can make people respond favourably to an ad and transfer that positive feeling to the product being advertised.

In an experiment, 12 press ads were shown to two groups. One group saw the ads on their own, the other group had music playing at the same time. The second group had a heightened emotional response to the ads. Although it may not be appropriate to play music during every business presentation, the general point about the tests (and my own experience) is that *it helps to use an emotional appeal.*

Music is not the only means. The words you use, the drama in your voice, the oratorical devices that Churchill, JFK and Martin Luther King used so effectively, all contribute to an emotional appeal. When people's emotions are engaged their response is intensified.

The decade's political success story

Rapid results are always impressive, and one of the most dramatic journeys from zero to hero was between 2004 and 2008. In 2004, the American Democratic Party held their Convention in Illinois, to endorse the Presidential campaign of John Terry. As candidate for the Senate seat in Illinois, a certain Barack Obama was invited to deliver the Keynote address. He was virtually unknown.

Obama delivered an electrifying speech, without notes. It was a presentation of his ideas, his style and his ability to rouse a large audience. Hillary Clinton and Jesse Jackson were there, rising to their feet, applauding enthusiastically.

Obama was elected to the Senate and, just four years later, was elected President of the United States, defeating Hillary Clinton on the way, and John McCain in the final. And all because of his ability to deliver a rousing speech that reached the hearts of his hearers.

That's the power of a great presentation.

There were four essential elements in the Barack Obama speeches, especially those in his Presidential campaign, four elements that I shall cover later in this book. They were:

1 Good content.
2 Oratorical devices.
3 Vocal authority.
4 Engagement of the audience.

All four can be developed and employed by you. All four are necessary if you want to persuade and if you want to acquire the voice of leadership.

The power of the spoken word

In his 2004 Keynote address, Obama engaged his audience by emphasising the importance of unity in America: 'There's not a *liberal* America and a *conservative* America. There's the *United States* of America.' He also placed himself alongside the man or woman in the street by telling his personal story, the 'unlikely' climb of 'a skinny kid with a funny name who believes that America has a place for him too.' When he spoke of 'the audacity of hope' he lifted the spirits of the average American.

That one speech, that single presentation, not only propelled Barack Obama into the Senate and later into the world's top job, it also *demonstrated the power of the spoken word*. It is that power which is available to you when you stand to make your case.

> *You will have many opportunities to practise the new skills you will learn from this book, because you make far more presentations than you might think. And every time you make one you have the chance to achieve great things.*

In simple terms, making a presentation means putting your point across in such a way that your listeners can understand, accept and be prepared to act on what you say. It can be formal or informal, but it usually has a business purpose, and will therefore have some focus and structure. It could be a new business pitch, a workshop or seminar, or it could be a speech.

In times of war

You already know of the impact Winston Churchill made with his radio broadcasts during the Second World War, rallying his nation when they were almost on their knees. You may also have heard of the *Gettysburg Address*, a two-minute declaration by Abraham Lincoln, to honour the thousands who had died in bloody battle during the American Civil War. That two-minute speech is probably the most famous speech in American history, and the one most quoted. More recently, another military leader made a worldwide impact with a short impromptu speech to his troops on the eve of battle. This is what he said:

> *'We go to liberate, not to conquer. We will not fly our flags in their country. Iraq is steeped in history. It is the site of the Garden of Eden, of the Great Flood and the birthplace of Abraham. Tread lightly there.'*

Those were the words of Lt Colonel Tim Collins during the Gulf War in 2003. He was the 42-year-old commander of the Royal Irish Rangers, and he gave a stirring speech to his troops just hours before the battle to liberate Kuwait.

There were two dramatic results. First, the Royal Irish Rangers were inspired to fight with valour and distinction. Second, Tim Collins became an internationally acclaimed hero and celebrity. He was immediately promoted to full Colonel and later offered command of the SAS and SBS, a promotion which carried the rank of General.

All because of that one speech.

Essential tip

▧   The right language can motivate people to do amazing things.

How does it apply to you?

Now, you may say that Tim Collins, Barack Obama, Winston Churchill and Abraham Lincoln are examples of global leaders, and you may be thinking, 'I'm not in that league. How does it apply to me?' So I had a look at some small businesses started by inventors and entrepreneurs, and the first thing that struck me was how much they benefited from guidance on turning their ideas into successful businesses.

The information that helped them was always available, in books, on the internet, and in many other places. But what made the difference to them was sitting in a room, listening to a live presenter who fed them the information in a structured way, explained its significance and showed them how to adopt and adapt it to suit their own circumstances. Most importantly, they were often inspired by something that someone said, energised and made bold enough to do something different.

Let's consider the value of making a good presentation, and how you can *recognise the success it brings*.

The word 'presentation' usually creates an impression of someone standing in front of a screen, talking about the words and pictures on that screen, and possibly even reading out the words that everyone can see for themselves. In the next chapter I'll explain why this does not work, but for now let me stress that it's the wrong image to call to mind.

Let's go back to basics. Why would you make a presentation?

You may think it is to inform, but surely the reason is to persuade your listeners to accept your proposition, your reasoning, your solution to some problem. For that to happen you have to be persuasive, and that's a two-way street. They have to like you and accept your line of reasoning, and you have to make it relevant to their needs or interests.

It's not enough to be entertaining or even impressive. That's not usually enough to get people to change their thinking, attitude or behaviour.

The Greeks valued oratory

I'm often reminded of two orators in ancient Greece, Aeschines and Demosthenes. Aeschines was a polished speaker, but a bit aloof, who nevertheless became one of the leaders of the Athenian independence movement against Philip of Macedon, father of Alexander the Great.

Demosthenes was an orphan who, when he came of age, decided to sue his guardians who had robbed him of his inheritance. But he had a pronounced stammer. So he learned what he could about law and overcame his stammer by filling his mouth with pebbles and jogging by the seaside. It helped him to develop a powerful speaking voice and style, and when he sued his guardians he won the case so impressively that people urged him to go into politics. He then rivalled Aeschines in the anti-Philip movement.

Here's the difference. When Aeschines spoke, people said, 'How well he spoke.' But when Demosthenes spoke, they said, 'Let's *march* against Philip of Macedon!' So which would you rather have, applause or action?

A presentation that results in the action you want is a success. The presentation that wins you applause has usually failed. When your listeners notice and admire your technique, when they tell you, 'That was a terrific presentation!', they have received a performance, nothing more.

When you watch a drama on the stage or screen, a good performance has you believing in the characters. You don't want to notice their acting techniques. So praise is good to have, but it is not the objective.

So, a last word on the need for presentations from David Ogilvy, founder of the advertising agency Ogilvy & Mather:

> *'In the modern world of business, it is useless to be a creative original thinker unless you can also sell what you create. Management cannot be expected to recognise a good idea unless it is presented to them by a good salesman.'*

In the next chapter we shall consider why presentations fail, before moving on to the thinking and techniques that will enable you to make winning business presentations, and also arm you to cope with the anxiety about public speaking.

Summary

- Demonstration works better than claiming
- Press the emotional hot buttons to engage your listeners
- Believe in the value you offer to others
- Don't underestimate the power of language
- Aim for action, not applause
- You must be able to sell your great ideas

Why presentations fail

In this chapter

Here are three startling research findings:

1 Eighty per cent of business presentations fail.
2 A survey of CEOs in Florida revealed that over 71 per cent of them admitted to dropping off during business presentations.
3 A Practical Training for Professionals (PTP) survey revealed that 75 per cent of UK managers find business presentations boring.

In my own experience of working with business people on their presentations since 1994, at least 80 per cent of them were somewhere between poor and embarrassingly bad.

The actual percentages are not important. What matters is that the time and expense invested in preparing and delivering presentations is, most of the time, wasted. You can add the indirect cost of diminished reputation and business lost simply because the presentation failed to impress. If you could quantify that cost it would be a frightening figure.

> *If a business presentation's purpose is to obtain business, it must be persuasive and possibly even impressive. If it is boring, if it is poor, if it is inferior in any way, not only will it fail to win the business, but it will create a negative impression of the presenter.*

Presentations tend to fail because certain disciplines have not been followed. The necessary techniques are relatively easy to acquire. What is more important is to understand *the process of persuasion*. It's the way to get more positive results from speaking in public.

Does it matter if a presentation fails?

Apart from the costs mentioned above, let me tell you about the contribution to the catastrophic failure of space shuttle *Columbia* made by a poor presentation slide. In a presentation about damage in a previous incident, important information about tile damage was hard to read on a cluttered slide and got missed.

Ambiguous language played its part too, and caused people to pay little attention to this phrase: *'Test results do show that it is possible at sufficient mass and velocity.'*

The word 'it' actually refers to 'damage to the protective tiles'. On the same slide, the words 'significant' and 'significantly' appear *five times*, thus losing their power to attract attention. That's a vital clue to the use of language in a presentation.

As a result of the damage during launch, when *Columbia* re-entered the Earth's atmosphere, a briefcase-sized piece of insulation broke off the main propellant tank, striking the leading edge of the left wing and damaging the shuttle's thermal protection system. *Columbia* disintegrated over Texas, and all seven crew members died.

The *Columbia* experience proves that it's not enough to put in the essential information. It must be done in such a way that it will be *received and understood*, because people do not pay close attention at all times, and they cannot be expected to extract the vital information themselves.

As the presenter, you will always have to *point out the meaning of the facts* you include, not least because they might mean different things to different people.

Essential tip

■ It's not enough to include the right information. You must explain it.

Check how well you are presenting

Presentations fail because certain essentials are not carried out. Like any business activity, a presentation requires certain disciplines to be observed. Why not *check now* to see if your own presentations are at risk?

Here's a checklist of 30 important considerations in preparing and delivering business presentations. In the 60 seconds or so that it will take you to answer the 30 questions, you will gain a clear understanding of your chances of success.

Just answer each question by ringing either YES or NO, then add up all your NOes at the bottom.

Essential one-minute presentation checklist

Why should people listen to you?

1	Do you have a 15-second Elevator Speech?	YES/NO
2	Are you an acknowledged expert in your subject?	YES/NO
3	Do you always speak with passion and conviction from the platform?	YES/NO
4	Is your speech or presentation always focused on making some change?	YES/NO
5	Can you stand and 'say a few words' at a moment's notice?	YES/NO

What makes you a 'must have'?

6	Can you state your Unique Selling Proposition in ten words or less?	YES/NO
7	Can you list five things about you or your business that distinguish you from competitors?	YES/NO
8	Would you pay to hear you speak?	YES/NO
9	Do you always get good, positive feedback when you speak or present?	YES/NO
10	Do you make your proposition in a compelling, dramatic way?	YES/NO

Preparation

11	Do you follow a well-defined structure that helps listeners stay on track with you?	YES/NO
12	Can you go from blank page to first draft in 15 minutes?	YES/NO
13	Have you recently had coaching or training in speaking or presentation skills?	YES/NO
14	Do you open with a memorable hook?	YES/NO
15	Can you easily speak the sentences you write in your script?	YES/NO

Using visual aids

16	Can you present without PowerPoint?	YES/NO
17	Do your slides carry a maximum of 30 words?	YES/NO
18	Do you know how to blank the screen with a single key?	YES/NO
19	Can you skip to any slide without scrolling?	YES/NO
20	Do you present without looking at the screen behind you?	YES/NO

Being persuasive

21	Do you follow the AIDA sequence of persuasion (see p. 84)?	YES/NO
22	Do you identify and address the needs of your audience?	YES/NO
23	Do you usually get business as a result of your presentations?	YES/NO
24	Do you tell your audience what to do at the end of your speech or presentation?	YES/NO
25	Do you usually get questions when you have finished speaking?	YES/NO

Connecting with the audience

26	Do you like the way you look when presenting?	YES/NO
27	Do you like the way you sound when presenting?	YES/NO
28	Do you ask rhetorical questions during your speech or presentation?	YES/NO
29	Do you use oratorical devices such as triads and anaphora?	YES/NO
30	Do you usually get a good connection with the audience, with hearty applause?	YES/NO

TOTAL:

How did you score? Almost no one gets away without a few NOes, and most people say the checklist reminds them of things they know they should be doing but usually forget. The NOes you mark will tell you why your own presentations are at risk.

Two common misconceptions

In a moment I'll deal with twelve common elements that make a presentation fall short of expectations, but first let me dispense with a couple of *misconceptions* that seem to have taken root.

The first is the astonishing and frankly incredible claim that speaking in public is a person's greatest fear. It is claimed that this activity outranks the fear of death!

The American humorist Jerry Seinfeld puts this into context. He says that if this were so, then the average person delivering the eulogy at a funeral would rather be in the box! That is clearly nonsense. But it will get in the way and cause failure if you believe it.

Of course there will usually be some anxiety about presenting, but it is perfectly natural to feel nervous when the spotlight is turned on you, and this book will help you deal with that nervousness.

The second misconception is that a presentation consists of simply telling your stuff. Most business presentations fall into this trap. They consist of a linear description of the product or service being offered, following this kind of sequence:

- this is who we are;
- this is what we do;
- we did it for these people;
- we'd like to do it for you.

It's one-way traffic. The presentation is prepared, complete with slides, and the presenter delivers it, hoping for a positive response.

An example of this approach was when Alan Johnson was appointed Britain's Shadow Chancellor of the Exchequer in 2010. He called a press conference at the House of Commons, turned up and read out a speech that he clearly had not written himself, took no questions, and left.

Not only that, he actually gave the journalists a copy of his prepared text!

Why, then, was it necessary for him to turn up and read the text to them? Wouldn't it have been more efficient to send it as an email? And yet, that is exactly how many business presentations are delivered. The text is prepared, with accompanying slides, and the CEO turns up to read the script and press the button to move the slides along.

What does your presentation actually cost?

Is it any wonder that people switch off and even drop off during such presentations? Try this: add up the cost of all the hours you and others spend preparing a presentation. Then add what it costs for your audience to be there – the total of their earnings per hour.

That large figure is the minimum level of value that you must deliver. If you bore or irritate them, you can add the hidden cost of lost goodwill plus the business you could have secured with a successful presentation.

A simple calculation like that will put into perspective the value in developing the skill of preparing and delivering a presentation that works. That skill has been devalued by the commonplace term, 'presentation skills', because there are so many unqualified people offering presentation skills training. Just as dangerous is the view that it was enough to have attended a one-day training programme on presentation skills in 1982.

Every business presentation is a 'sales pitch'. That is not to say you have to be in sales or have a product or service to sell, but your presentation expresses *your point of view*, and you will want your listeners to accept, agree with and perhaps even admire what you say.

It requires the persuasive skills of a professional sales person, someone who is in tune with what turns people on these days.

The Dirty Dozen

Let me now turn to some of the common failings of business presentations. I have compiled my own Dirty Dozen, and they fall within three categories:

A Content.

B Style.

C Delivery.

A Content

1 Too much

Your purpose must be to inform and inspire your listeners to accept your proposition. They do not need to know the history of your company or the CV of its founders. To use a sporting analogy, if you were the owner of a football club, and someone wanted to interest you in 'buying' a new striker, what would you want to know? Would you want to know how many GCSEs he got at school, the size of his boots, how hard he kicks the ball, how many hours a week he spends on fitness training? No. You'd want to know how good he is at scoring goals.

The same applies to your audience. They want to know what directly concerns them, the benefit it can deliver for them, and they want it without the clutter of excessive supporting material.

A related caution is about the number of slides. Don't have a slide for everything you say, or you might as well use a video. Guy Kawasaki, formerly marketing chief at Apple, uses only ten slides, one for each of his top ten points.

2 Not enough

Some presentations go to the other extreme, and leave out essential content. Your case must always be complete. It's not safe to make assumptions about the common ground between you and your audience, any more than to assume they know nothing. A little research beforehand should give you an indication of how much they are likely to know.

It is equally important to *repeat the essential elements* within the presentation itself, because your listeners may have missed a link between two or more of your points.

Think about road signs. Isn't it often the way that the crucial final turning is not marked, perhaps because it is assumed that it is known to all local drivers? Have you ever been given directions to an unfamiliar destination far from home? People almost always leave out some vital piece of information, and you have to call for the missing bit.

3 No common terms of reference

When you are telling people things that are new to them, you must make it easy for them to understand and relate to your information. It is familiar to you but may be foreign to them. So always start with what they already know, and give them a set of reference terms that will help them stay on track with you, but *avoid the use of jargon and acronyms*.

When I went to work at *Reader's Digest*, London, I was totally bewildered during my first creative briefing. Everyone was using acronyms, to show that they were on the inside track. It gave them comfort, but it did nothing for me.

I once attended a presentation about heating. The presenter used terms like BTUs (British Thermal Units), but did not tell us how to understand the relationship between the source of heat and the cubic capacity of the room. The result was that we were expected to absorb and understand a parade of figures, and all the laypeople quickly switched off.

Terms of reference also define the context and scope of your presentation, so that your listeners know what to expect.

4 Lack of structure

One of the most common errors is the absence of structure. There is a common misconception that a linear account is a structured presentation. That could be, for example, a chronological sequence: this happened first, then that, and this came next.

Three minutes into your presentation, your listeners will be thinking, 'Where are we going with this?' In fact, if that question ever arises, either aloud or in the mind of a listener, you need a better structure.

A structure has two main purposes in delivery, and one in preparation. In preparation it directs your content and focus, while in delivery its purposes are: (a) to keep you on track, and (b) to help your listeners follow you.

There are some simple patterns to guide you, which are covered in Chapter 5. They will make a considerable difference to the effectiveness of your presentations.

5 All tell, no sell

I referred to this earlier in this chapter, but let me expand on it a little, because it is an error that some people cannot see if they have what Tom Peters called 'the engineer's mentality'. According to him, such people believe that the facts speak for themselves, and that 'truth and virtue will automatically be their own reward'.

If that were so, it would be sufficient to write the facts on one side of a small piece of paper and send it to those you wish to do business with. If it were so there would be no need for marketing, for sales teams, for advertising.

The reality is that *facts tell but feelings sell*. We all buy on emotion and justify with reason. A presentation that is full of data will do no more than inform. We need to interpret the data and say what we want our listeners to understand and feel about it.

I have come across some people whose approach has been: 'This is my product, here is the specification; you can see it is good, so you must buy it.' If that sounds ridiculous, you'll understand the need for more than facts alone.

Essential tip

■ Make your content easy to take in and understand.

B Style

6 Over-full slides

Slides and other visual aids are meant to support but not supplant you, the presenter. Some people load their slides with all the things they want to say, not just the things they want to talk about. The slides become the script on the screen. Others limit themselves to three or four sentences per slide and don't understand why even that is too much.

It's very simple: we have all been conditioned to read sentences, so if you show a sentence on the screen your audience will read it. While they are reading it they are not listening to you. People who knock bullet points do not understand how slides work in support of the spoken word.

You should be able to look at a slide and take in its meaning in an instant, recognising it as *a legitimate summary* of what is being said. The bullet point's function is the mechanical one of signalling the point that is being made by the presenter. In Japan, Masayoshi Takahashi created the *Takahashi Method*: single words or short phrases in very large letters, that you can instantly understand. Like this:

7 Cluttered slide design

It is tempting to give slides fancy backgrounds, to use multiple colours and a range of fonts, even on the same slide.

At the less well-informed end of the scale, there is type that is too small to read, type that is unsuitable for presentations, and type that is just ugly.

Here is an example of type that is too small (It's actually 12 point!)

HERE IS AN EXAMPLE OF UNSUITABLE TYPE

Make up your own mind about this one

Too much content can create a slide that hinders rather than helps the presentation, like this one:

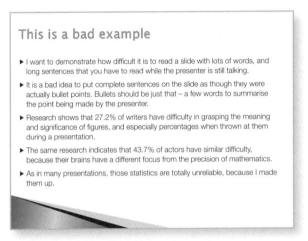

8 *Wrong language*

If you write a script, remember that the language that is written to be read is not the same as the language that is written to be said. Also, Peggy Noonan (speechwriter for presidents) recommends that 'you must be able to say the sentences you write'.

Avoid business buzzwords. Some people think it signals that the presenter is 'in the know', uses the right jargon, so must be in the right stream. The opposite is true. The presenter will be dismissed as pompous and unclear.

Grammatical errors will also get the presenter an uphill climb, especially phrases like 'Comprises of' or 'As a valued customer we'd like to offer you . . .', or 'your' instead of 'you're', 'it's' instead of 'its' and 'there' instead of 'their'. These howlers appear time and again on slides as well as websites and promotional literature.

Another caution is about those words that mean different things to different people. There are many examples of the differences between UK English and US English. In the US, to 'table' something means to put it on the back burner. In the UK it means the opposite – to present it.

Another example: an American firm was negotiating to supply to a Japanese firm. As they were going through the final checklist of agreed points, the American said 'No problem' to each point in turn. Unfortunately, the Japanese heard, 'No, problem.'

Those whose first language is not English don't always understand negative terms to signify positive intentions.

Essential tip

▓ Let your visual aids support but not supplant you.

C Delivery

9 Reading the slides

Just about the worst thing a presenter can do is read out the words on the slide, as though the audience had never learned to read! It is the error that will alienate an audience most quickly and guarantee the presenter some harsh criticism.

I was once asked to give a talk on presentations to the sixth formers at a major public school. They were learning to use PowerPoint and, following the (bad) examples of their teachers, believed it was the right thing to do, to read from the screen. I asked them: 'Who was the first person to read to you?' One of them said, 'My mum.' And what time of day was that? He said, 'Bedtime.' And for what reason? 'To send me to sleep.' Why, I asked, why would you want to do that to your audience?!

The presentation that is delivered in this way, with the presenter reading out the words on the screen, will usually come with *a cast-iron guarantee of failure.*

10 Boring voice

The content may be good, the slides well-designed and slickly executed, but if the presenter sounds boring, the presentation has little or no chance of success. Radio and television have conditioned us to prefer interesting voices. We know how the person should sound, whether they be reading the news or starring in some drama.

At the same time, the massive numbers of messages that assail us daily have trained us to switch off. We have a very low threshold of boredom. There is no excuse for sounding boring. If you can't be excited about your proposition, why should your listeners care?

11 Self-centred

Some presentations come across as self-indulgent and focused on the presenter. This can manifest itself in the content, style or delivery. For example, it may be full of opinion without supporting fact. It may be a celebration of the presenter's credentials or their company's position in the marketplace. A single instance can turn off the listeners.

A common example is the opening sentence that goes, 'We at XYZ Company . . .' That phrase signals self-congratulation, and people do not attend business presentations to hear that. It is a mistaken belief that audiences will pay attention only if they have first been impressed by the credentials of the presenter or their company. The reality is that they are only interested in what you can do for them.

Never forget that they are permanently tuned to WII FM – What's In It For Me?

12 Going on too long

I once attended a dinner at a sports club, to which a famous athlete had been invited as the after-dinner speaker. The club steward decided to introduce him. He reminisced about his own running background and (limited) achievements. He told us how he had followed the guest of honour's sporting career. He gave an extended history of our club, listing some of the luminaries who had represented us with distinction in the past. He went on for 40 minutes, by which time the guest speaker must have lost the will to live. The rest of us certainly had.

It is rare that people complain about a presentation being too short. But they *always* complain if it goes on too long. In conversation, anyone who speaks for three minutes without interruption is boring. Three minutes! We endure longer presentations because we dip in and out, but if the presenter is boring, we stay switched off and pray for the end to come swiftly.

Essential tip

■ Make the effort to sound interesting or your listeners will switch off.

There are certain exceptions. Some presentations are instructional and may require complex slides with charts and other detail that needs careful explanation. In such cases, you will have to refer to the slide on the screen and even sometimes read out what's written there. But these are exceptions which will not apply to most business presentations.

In fact, it's probably a better idea to put such complex matters in a handout. It's really not a good idea to try and explain complex detail in a business presentation. Far better to talk about the conclusions, or the consequences, of the research or whatever else you were going to put on the slide.

There are, of course, quite a few other errors that can impede a presentation, but the most important one to remember is not knowing your stuff. Or, to express it positively, be sure to know your stuff.

Know your subject, know what you want to tell your audience, know why they should care, know what questions they are likely to ask you, and know the answers.

Do you need PowerPoint? Not necessarily. You should be able to present with no visual aids, or with just a flip chart. We'll address this in Chapter 5, Drafting your presentation. But first, let's consider what goes on before you start preparing your presentation.

The next chapter will help to clarify your thinking about the purpose of a business presentation, and the role that you will play as presenter.

Summary

- Failed presentations can have drastic consequences
- Use the checklist now (and regularly) to cover the essentials and avoid the errors or omissions that cause failure
- Don't just deliver your content. Help listeners to 'get it'
- Visual aids should help, not hinder
- Know your stuff

Essential planning: getting started

In this chapter

- The importance of your role as presenter

- Beware relying on facts alone

- The 4 Ps of presenting: Principles, Purpose, Planning and Pre-qualifying

- Five key questions to focus your mind on your purpose

- How to fact-find

- Elevator Speech shortcut

Your business presentations will usually be about some aspect of your business, so they may be regarded as a marketing activity, and need to fit with your company's marketing objectives. *Marketing* has endless definitions, but essentially it is *the process of discovering what people will buy and offering it to them* on terms they find acceptable.

A presentation is one of the means of delivering that offering. It has a bearing on the way the company is perceived in the market place, even if it is about some activity that seems far removed from marketing.

Obviously a business presentation cannot be considered in isolation. Being part of the marketing plan it must fit with the disciplines of that plan. Even before getting down to preparing a presentation we need to think about the people it is intended for, and the reasons why they should be interested in what we have to present.

The more you think about that, the more obvious it will become that *you need to be persuasive*, and that means appealing to people's emotions. It has been well documented that we buy on emotion and justify with reason. So it's not enough just to state your proposition. You have to do what it takes to get your point of view accepted. That requires persuasion. And persuasion follows a particular path.

Before we go into how that works, let's consider the part that you play, as the presenter.

The importance of your own role

A business presentation is a sales pitch, pure and simple. You have a proposition to put before a chosen audience, and you want them to listen, accept and act. It could be a plan, a strategy, a piece of research, a project, an event, a charity sponsorship . . . anything at all.

The important question is, why are you making the presentation? If it's purely to inform, why not send the information by email? If, on the other hand, you want your audience to accept what you present and make a decision about it, you will be aiming to persuade, and that means making a pitch.

You need to understand the vital part that you play, quite apart from the content and the visual aids that you may employ. You, the presenter, are the key component of any presentation, and it will be essential to get your own thinking right if your presentation is to succeed.

Whether it is an 'in house' presentation delivered to a single client, or an 'open' presentation delivered to a self-selecting audience, as in a seminar, conference or other general event, the rules tend to be the same. While there may be some differences in approach, your role is always paramount, as the expert on the topic being presented.

Elevator Speech

A very good starting point is to identify why it is *you* making the presentation. What are your credentials? Do you have an *Elevator Speech*?

You should be able to state, in 15 seconds, why someone should be interested in doing business with you, or at least in finding out more about you. You should be able to say what your personal added value is, and why your audience should be receiving the presentation from you, and not some other person.

You need to be sure that the presentation derives from your own expertise, and feel confident that you could field questions and explain your personal point of view on the topic. When you can do that, you will have confidence in the message of your presentation.

To develop your own Elevator Speech, just answer these questions:

1 Can you name one or two problems (in your field) that others can identify with and which make them nod their heads in agreement?

2 What is the consequence, to them, of that problem or those problems?

3 How do you provide the solutions?

Your objective is to get the other person to say, 'Tell me more.' Just that.

> ## Essential tip
>
> ▨ You need to identify your own added value.

Beware relying on facts alone

In the previous chapter I mentioned the 'engineer's mentality', a concept put forward by Tom Peters, the American business guru who co-authored *In Search of Excellence*. It refers to a belief that the facts speak for themselves, and that 'truth' will automatically be its own reward, as though there were no need for any of us to intervene or help the process of getting that truth accepted. If that were so, there would never be any need for sales people or advertising.

Think about your own experience of reading about some topic that may be new to you. It's all there, in the book or handbook, but don't you sometimes wish you had an expert there, telling you what it all means, and explaining how it works? A lot of people say they have that experience every time they assemble flat-pack furniture!

Clearly the facts do not speak for themselves, and we need someone to explain them to us in terms that we understand. I recently had a meeting with a bright young man with a specialist knowledge within the financial sector – a topic which many might consider arid. He took the view that, when he attends a conference on his subject, he is there to learn something new and doesn't care how it is packaged or delivered. By claiming that he was interested only in the factual content of a presentation, he seemed to be endorsing the engineer's mentality and he would not, at first, admit to any emotional response to the content of a presentation, or to being interested in the presenter beyond their credentials.

As we discussed his typical response to new information, he eventually admitted to a feeling of 'satisfaction' at receiving and recognising some new angle on his subject, because his own market value depended on his knowledge, and on being up to date.

I put it to him that his satisfaction came from realising the *potential* in the new angle, not from the facts alone, and he agreed. If it caused him to open his eyes a little more and think to himself, '*That's* interesting!' he was responding to more than the information alone. He was responding to the *possibilities* in the new information. The 'satisfaction' at realising those possibilities and how they added to his own fund of specialist knowledge, constituted an emotional response. He accepted that.

Getting that agreement was not easy because I noticed that he is not easily engaged. He gets lost in his own thoughts, and his attention is intermittent. He probably listens to presentations in the same way, not fully engaged. It's quite common for listeners to drift off, which is why, as a presenter, you must strive to engage them, and create in them a realisation of the possibilities in the information you impart.

Essential tip

■ Instead of conveying facts, give the audience your 'take' on the facts.

To get you started on the right track, we should remind ourselves of the fundamentals, the **4 Ps of presenting**:

1 Principles.

2 Purpose.

3 Planning.

4 Pre-qualifying.

Principles

Let's return to the question that is important to ask at the start: why do you make a presentation?

Whenever I ask that question in the training room, someone always says, 'To communicate some information.' I then suggest that it would be more efficient in an email, which could be read at a time that suits the recipient. Everyone's time would be saved, and it would be much less expensive than a formal business presentation. There must be some greater purpose, and that would be the first general principle.

So the first general principle is: *aim to bring about some change* – in the thinking, attitude or behaviour of the audience.

If you do not have a change in mind, don't make the presentation, unless you are happy to treat your presentation as an entertainment. You need to determine at the start what your objective is, and what action you want your listeners to take when you have finished.

It goes without saying that if you do not have a change in mind at the start, then it is highly unlikely that any change will result from your presentation. It will be as though you were like a stick in a bucket of water. When the stick is removed, there are ripples for a time, but when they subside, there is no evidence the stick was ever there.

The second general principle is: *filter the facts*.

Facts and figures are neutral. They require interpretation and context to become Information. For example, consider a table that measures 6ft × 4ft. Is it a large table or a small one? If you wanted to fit it into an alcove that's 3ft wide, you'd say it was too big. But if you wanted to use it as a dining table for 20 people, you'd say it was too small.

Similarly, if I told you that a certain person could run half a mile in two minutes and ten seconds, you'd need to know the age and gender of that person, typical performances and the record for the half mile in that group, before deciding if that was a good performance or an average one. For an international level event it's a poor performance. But for 13-year-old girls it would be fantastic!

Information needs to be understood if it is to become knowledge. But that knowledge needs to be internalised and passed through your personal filter to become your point of view, your wisdom. So ask yourself:

- What do you think about the facts?
- What do you want your audience to think about the facts?
- What do you want them to do with the facts?

The third general principle is: *communication requires a receiver* who is able to understand what is being said.

Picture yourself alone on a desert island. No communication is possible because you have no one to receive what you say. Now imagine a new castaway, someone who doesn't speak or understand your language, nor you theirs. This person can receive what you say, but not understand it. How effective will your communication be? You therefore need to frame your content in terms that are easily understood by your audience.

Purpose

In Chapter 1 I mentioned my time at the *Daily Express*, and the effect of changing the approach of the advertising sales team. I gave them five questions to ask themselves before every sales appointment, and said, 'You must have the answers to those five questions on the tip of your tongue *before* going in to see a prospect.'

The same five questions will help your focus when preparing a presentation. They are:

Five key questions for focus

1 Why am I here?

2 Why should they see/hear me?

3 What can I offer that they cannot get from someone else?

4 What do I want at the end of the meeting/discussion?

5 What's the least I will settle for?

1 Why am I here?

It's not enough to say, 'Because someone arranged it.' Why have you set up the presentation? Is it to promote some new idea, is it part of a rolling campaign, is it to make a sale or to create an awareness, is it to create advocates among influencers? Have a purpose that's more specific than just to throw mud at the wall and hope that some will stick.

2 Why should they see/hear me?

Just imagine arranging a presentation for a large corporation and receiving a phone call from them to ask for a postponement. As you start to match diaries for a new date, the other person says, 'Just remind me, what will the presentation be about?' If you do not have a compelling reply on the tip of your tongue, you'll be blown out of the water. You should also be very clear about why they need to hear it from *you*, and not from anyone else.

3 What can I offer that's different?

This is about your USP (Unique Selling Proposition) and more besides. You must know what is different and special about your offering, even if it's about an idea, not a product, or a new way of looking at some well-established process, a new twist on a known idea. It should form the core message of your presentation – the one thing you want your listeners to receive and remember.

4 What do I want at the end?

You must have a clear idea of what you want to happen when you have finished speaking. Never assume that your listeners will rush at you with open chequebooks to buy your product or service.

However much they may be impressed with your offering and what it will do for them, most people like to be asked. Tell them what you want them to do, and make it easy for them to comply.

5 What's the least I'll settle for?

Few people get a 100 per cent acceptance of their propositions, so be prepared for rejection, and have a plan B. That might be something as simple as agreement for some follow-up activity. You need something that lets you come away with your tail up. If you have planned your presentation properly, you should have agreement on the problem and on your solution. The least you should come away with should be an understanding of why your solution is not seen as strong enough to cancel out the problem or remove the pain.

Essential tip

■ Clarify, in your own mind, exactly why you are making the presentation.

Planning

You know what you want to give a presentation about, you have a date, a time and a venue arranged, and you have some idea of the people who will be in the audience. Or do you? It's very important to have a clear idea of the kind of audience you will be addressing, even if you have to make an educated guess about them (see Pre-qualifying below). The content of your presentation, the tone and the language you use, could all derive from what you know and feel about the audience.

Here are some questions you should ask about them in advance:

■ How many will be present?
■ Who are they (job titles, level of seniority, gender mix)?
■ Are they likely to be friendly or resistant?
■ Do they already have a position on your topic? What is it?
■ What is their current approach to the problem you can solve?
■ What is the pain that you can remove?
■ Are there any cultural or educational issues you should consider?
■ What are their expectations?

Pre-qualifying

There are two basic rules about making a pitch, which will save you a lot of heartache and avoid wasted effort – if you follow them.

The first comes from the field of *direct marketing*. It is this: don't try to convert non-users into users, just concentrate on converting users to your brand.

The second comes from *direct selling*: make your pitch only when you are addressing decision makers or decision influencers.

I learned the *first* when I worked for *Reader's Digest*, one of the most skilled practitioners of the art of selling in print. In their huge database, in addition to names, addresses and purchasing history, there was information on the known preferences for the various products in our portfolio. Our testing programme (pre-qualifying) was designed to discover more about those preferences, to identify those people who were most likely to buy the product we were about to offer.

For example, the mailing for a music collection went only to those who were known to be interested in music . . . of that kind. It reduced wastage, generated high levels of acceptance, and updated the known preferences. And it was entirely consistent with the marketing principle: find out what people want and offer it.

Trying to convert non-users into users is a slow, expensive process, with very small returns. It may work some of the time, but is best avoided.

The *second* basic rule is just common sense. Unsuccessful salespeople often launch into their pitch to people who do not have the power to buy. In my very first job, fresh out of university, I was the most junior person in the advertising department of an industrial company. One day a salesman asked to see the head of my department, and I was sent to get rid of him. I found his product fascinating and asked lots of questions about it. He gave me the full presentation, although I was clearly too junior to make any buying decision. I gained some new knowledge, but he just wasted his time.

His focus was wrong. He must have known I was not the decision maker, yet he 'hoped' I would transmit his message to the real decision maker. In other words, he was prepared to rely on someone less knowledgeable about his product and less skilled in the process of persuasion, to do his job for him.

The exception is when you are faced with *genuine influencers*, people with the power or responsibility to select and nominate the chosen few who will have the chance to pitch to the decision makers. In such cases, it is fine to make the pitch, but only if you also determine the rules of engagement, i.e. what happens next, and the basis on which the selection will be made.

For *'in house' presentations*, whether one-to-one or to a group, always ask how decisions are made on the product, service or whatever else you are offering, and whether it would be possible to include all those involved in such decisions. Much depends on how you set up the appointment. If you are open about your purpose, and its likely benefit to the prospective customer, you are far more likely to present to the right people.

For *'open' presentations*, always include some qualifying statements or questions in your promotional material, such as, 'Intended for people who buy (product or service)' or 'If you have ever wanted a safer way to avoid wasting time and money on the wrong (product or service), this presentation provides the answer.' For seminars, you may care to consider sending out a pre-qualifying questionnaire which asks about their expectations and makes clear what your own objective happens to be.

Pre-qualifying is one of the important tasks to undertake when planning a business presentation. It includes asking the right questions, partly for your own sake, and partly to condition the thinking of your prospects.

Essential tip

■ Unless you present only to those who can make or influence decisions, your presentation will only be a performance.

Four kinds of questions

Fact finding

Every sales manual tells you to start by asking questions to 'fact find'. But unfortunately that prompts most people to ask factual questions: how long have you been in business, how many staff, current position, etc. Factual questions can be irritating, so keep them to a minimum, and relevant.

Move on quickly to the important persuasive questions, about:

■ problems;

■ effects;

■ importance.

About problems

Be direct. After all, you are there to try and solve problems.

Examples:

- What are the main issues you have with this type of product?
- Have you ever been let down by your current supplier?
- Is there information you'd like to have, but which you are not getting now?
- What would simplify your procedures?
- Do you need to cut costs/save time/extend the application?
- Some of our customers have had reliability problems with this (competitive) product in the past. Has that been your experience?

About effects

What are the consequences and knock-on effects of the problems you have persuaded your client to reveal?

Examples:

- Is this slowing down your overall rate of production? By how much?
- What is the effect of a delay in delivery?
- If the system is dependent on Ernie's knowledge, what happens when he's on holiday or off sick?
- How long before this glitch leads to a full breakdown of the system?
- Will this lead to job losses and budget cut-backs?

About importance

Find out how important it is to correct the problem – to the company and to the decision maker. Focus on the cost of not correcting it.

Examples:

- Have you worked out what this problem is costing you?
- How urgent is it to put it right?
- What is it costing you to have unreliable deliveries?
- How much market share do you think you are losing because you don't have up-to-date information?
- How would it be if you could have [your corrective benefit]?

You must aim to get your prospect to admit that the problem is serious and costly and that they want it corrected NOW.

Essential shortcut to your Elevator Speech

Write the following three things, with a 2in/5cm gap between them, then add what is indicated by the italics:

1 You know how (*add a problem or two that people can relate to*).

2 Which means that (*write the consequence of the problem/s*).

3 What I do is (*state how you provide the solution*).

The next chapter deals with the content, and deciding what to say.

Summary

- The Elevator Speech helps identify the added value that you provide
- Deliver your personal 'take' on the facts
- Follow the 4 Ps of presenting: Principles, Purpose, Planning, Pre-qualifying
- Five key questions to remind you of your purpose
- Use fact-finding questions to uncover where the real 'pain' lies

4

Deciding what to say

In this chapter

- Choosing relevant content: Purpose, Audience, Topic

- The five main types of presentation: Entertaining, Informative, Analytical, Problem Solving and Persuasive

- The negative and positive elements of a presentation

- Talking about what you know

- Your role as problem solver

- Projecting your brand values

When you come to decide on the content of your presentation, follow the simple PAT formula: *Purpose, Audience* and *Topic.* Just remember, this chapter is about content, not about structure. That will be covered in Chapter 5.

Purpose: Why are you giving the presentation, and what are you hoping to achieve through it? What is the change in thinking, attitude or behaviour that you want to bring about? How strongly do you feel about that? If you find yourself saying, 'I've got to give this presentation and I don't know what to say', you should take a big step back and re-think your purpose.

One of the main reasons why presentations fail is that those delivering them are not engaged. The presentations are part of their routine responsibilities, and there is no conviction in what they say or in the way they put it across.

Quite a few such presentations were delivered by people in key roles within the Regional Development Agencies in Britain, which were set up to attract and promote business within their areas. In one example, there was a small department which had several million pounds available to lend to SMEs, small companies that were desperate for financial support. Yet the people with the money to lend prepared presentations that were truly boring, talking about the available financial aid as though it were a form-filling exercise.

I put it to them that they were in a position to make things happen, to help businesses grow and develop, to create new jobs, to bring increased prosperity to the region, even to save businesses from collapse through lack of funds. All exciting possibilities. Worth getting excited about! I said their presentations should not be about the *process* of applying for the funds but rather about the *opportunities* that the funding could create.

Your own presentations should arise *out of your personal conviction* that you have:

■ something to say;

■ something you want people to hear, not out of an obligation to fulfil an arrangement.

As I said before, presentations delivered without a commitment to the message will tend to fail.

Audience: Who will be there to listen to you, and why are they there? Many audiences contain people who attend because they have been instructed to be there, and may need to be engaged as early as possible. Others may have expectations that go beyond your brief, and you will need to rearrange those expectations.

■ What do you know (or can you find out) about their interests, demographics, motivation and relevance to your topic?

■ What do you feel about such people?

■ Are you in tune with them?

Every audience will be different, and your preparations should be re-tuned every time, no matter how often you give the presentation. Remember that there will almost always be some who think they know more about the subject than you. Think about how you would address them. Think about how you will engage them from the start.

Topic: What will you talk about? Start by writing the headline for an ad promoting your presentation. What's the main benefit you can promise? Make it as specific as possible, as though you were addressing a group of people who should be there, but who need convincing.

One of the most important people to convince is yourself. You need to identify the added value that *you* can provide through this presentation. That will give you your focus and raise your level of confidence.

Essential tip

■ When thinking about what to say, concentrate on how it will benefit *the audience*.

There are, of course, several kinds of presentations, and your approach will differ in each.

Five main types of presentation

For the most part you will be offering solutions to problems. No solution is of any interest or value unless first there is agreement on the problem and the need to put it right. Think about it: does it make sense to offer a solution before you know what the problem is (if there is one)?

Someone once told me *'Prescription without diagnosis is malpractice'*. How would you feel about a doctor who wrote out a prescription without first finding out what was wrong with you?

Equally, you and all your credentials are of no interest until you can demonstrate that you can provide value.

That's why it makes no sense to start your presentation with your credentials, or any claims of past greatness. Many presentations (and even expensively produced ads on TV and in the press) begin with something like, 'At [our company] we believe in such and such.' Frankly, who cares?

I'll return to this subject a little later, and in greater detail in the next chapter. For now, let's consider the five main kinds of presentations:

1 Entertaining.
2 Informative.
3 Analytical.
4 Problem solving.
5 Persuasive.

Entertaining presentations are designed to keep the audience amused. They include after-dinner speeches, and they have little purpose beyond entertainment and raising the profile of the speaker.

Informative presentations deliver new material to the audience, and may be reports on the progress of a project or about some new development or discovery in a subject that is familiar to the audience.

Analytical presentations deliver new thinking or an explanation of some current situation. They are frequently used in academic circles, but in business they provide the basis for decisions on some new direction. For them to be useful, they need to recommend some action to implement the new information.

Problem solving presentations are similar to analytical ones but contain, by definition, 'the next step'. Like analytical presentations, they can be simply academic unless they prompt some new action.

Persuasive presentations are the main concern of this book. They are the most common type of business presentation or motivational speech, designed to bring about some new thinking or behaviour, or to change a prevalent attitude. They are, therefore, likely to be challenging, and will need to follow the guidelines of persuasion.

Remember that a presentation is not the same as a conversation: you are an expert on the subject, with a specific message to impart. Know what that is.

What do you know?

Patricia Fripp CSP CPAE is a highly regarded professional speaker, based in San Francisco. As a teenager she went from the West of England to Hollywood, to be a hairdresser to the stars, and eventually set up her own hairdressing business, promoting it through talks she gave at organisations like Rotary, Kiwanis and Optimists. Just like you, when faced with a business presentation to prepare, she had to consider carefully what she would talk about.

She knew, instinctively, that she could not simply talk about her hairdressing business. She also knew that her audiences would resist an overt sales pitch. So she asked herself this question: 'What do you know that other people want to know?' So she talked about the *effects* of the work she did – the importance of appearance and customer service. Her hairdressing salon helped people look good and feel good about themselves, while her customer service made them feel acknowledged and special.

When I met her at a Convention of the National Speakers Association (NSA) in Texas, I asked her, 'What do people want from speakers like you?' She answered with one word, 'Wisdom.' She went on to explain that your wisdom is what people want to hear or learn from you. It is your 'take' on the subject, your unique point of view.

As we talked on, she said she was always being asked where she got the material for her speeches. She said she never runs out of things to talk about, because she constantly reviews her own life experiences, the high points and low points, the successes and the failures.

We all have experiences, and sometimes we get flashes of insight in the shower, in the car, on the beach. Those insights are gold dust. They represent new ways of looking at a familiar subject or problem, and they provide the filter for the factual content of your presentations.

Do you keep a journal? It's worth doing. Write something every day, especially on the topic on which you are likely to make presentations:

ideas, incidents, insights, interviews, interests. They could provide you with the original stories to illustrate some important points in your presentations. They will ensure that your preparation has already begun, when you decide to work on a new presentation.

Essential tip

■ You are the world's leading expert on . . . your point of view.

Relevance

If the subject matter of your presentation is not relevant to the audience, why should they listen to you? At the same time, you need to consider its relevance to your own objective and to your overall marketing plan.

The question of relevance can sometimes be quite complex. It is well known that there are secondary goals such as wanting to be perceived in a positive light, which may cause the presenter to modify what they really want to say. *Such goals are relevant to the presenter,* not necessarily to the audience, so a bridge has to be formed.

However, the starting point must be to ask if the subject matter is of interest to your audience.

■ Why is it of interest?

■ Does it relate to a pressing concern, some issue or problem that is causing them to lose sleep or money?

■ Will it show them how to gain some competitive advantage, or make savings in terms of either money or time?

To help you select from all the material you have on the topic of your presentation, here are a couple of prompts: think about what the audience wants to hear from you and, quite separately, what you want to tell them.

The audience want to hear you say three things:

■ something they already know;

■ something they didn't know before;

■ how they can use it.

Why would they want to hear *something they already know?*

Because they already know something about the subject of your presentation (otherwise why are you presenting about it?), it puts you both on the same track. It reaffirms their knowledge and helps them to accept you as someone who also knows their own subject. Also, it gets their heads nodding, either literally or metaphorically, and that's one of the first rules of selling: get a 'Yes' answer as soon as possible and you are on your way.

Moving on to *what they don't know*, this will be a combination of what they want and what you want. They want to hear something new, possibly startling, certainly positive, which will add to their own fund of knowledge and help to improve their performance in the market place. It has to be of *practical value*.

On the other hand, you want to present your own three things:

■ something negative;

■ a positive solution;

■ your own point of view.

Why something negative? *Because you want to bring about some change in their thinking, attitude or behaviour.* For them to accept the need for any change, they must first realise that something is not right. More than that, they need to be reminded of the consequences of that weakness, and of not putting it right.

Essential tip

■ To bring about change you must first make the audience believe things could be better.

Negative elements

Staying with the concept of putting things in groups of three, the negative half of the presentation will consist of these three elements (**SWE**):

1 Situation: how things are at present.

2 Weakness: what's wrong with it.

3 Effect: consequences of not correcting the weakness.

Situation

It's always wise to state the current situation, both to get on the same track as the audience, and to set the scene for pointing out where there is a weakness. Incidentally, the 'weakness' is not necessarily something negative. They may not be doing anything that's actually wrong, but it could be that they are not doing something that would benefit them more, and they may be losing their competitive edge as a result.

Of course it's no use simply giving a description of their business and their market, because that would be boring and you would lose their attention right away. It's always best to say something they didn't think you would know, and you'd get that by doing your research.

I once gave a talk to a group of dentists about building rapport with their patients. I showed a picture of a young boy with an apple on his head, playing the *William Tell Overture* at the same time, as I invited one of them to fire a toy crossbow at an apple on someone else's head. (No one volunteered.)

I said their relationships with patients were based on a similar level of trust, and showed them a picture of a terrified patient in a dentist's chair. They all roared with laughter at the familiar scene, recognising that they wield sharp instruments, like needles and drills, that could cause pain, yet they expect us to trust them not to injure us. They got the point (no pun intended!).

Weakness

It will always be important to research the business or the interests of the people you will be presenting to.

- The internet may be a good starting place to find out about their industry in general and perhaps their company in particular, through their website. If the presentation is for a prospect company, it's a good idea to ask them, in advance, about the issues that concern them.

- Another useful source of feedback is *customers*, i.e. people who already use the products or services of the prospect company or their competitors. Again, the internet might provide such information, if you search for comments on them.

- A third source is the activities and performance of others in the field, including *existing users* of your product or service, and possibly even colleagues within your own organisation. What are they doing that the prospect company has yet to adopt? Can you quantify the gain that existing users made by using your product or service?

If you are making an 'open' presentation, as in a seminar, you should still investigate the industry you will be addressing in your presentation, and identify where weaknesses lie in general. If, for example, your presentation is about telephone salesmanship, you could research:

- the conversion rates in call centres;
- the pet peeves of people who receive cold calls on the phone;
- the high attrition rates of call centre agents;
- examples of poor scripts.

Effect

Once you have identified the most significant weakness, you need to magnify its effect, and demonstrate the cost of not putting it right. Don't allow your listeners to say, 'We know things are not perfect, but nothing ever is, so we'll just live with it.'

As part of a campaign to save energy, I was giving a speech to a group of business leaders in the City of London about the need to reduce their carbon footprint. They all agreed that there was wasted energy in their firms, but were not sufficiently motivated to make changes. I pulled out my wallet and said, 'What if you could see the cost of wasted electricity every time lights are left on all night in an empty office? (I threw some money on the floor.) What if you could see the cost of your vans driving around with dirty engines? (More money on the floor.)' As I listed several sources of wasted energy, I threw more cash on the floor.

I continued talking, but I could see they weren't listening, so I stopped and asked them, 'What would you like me to do?' They replied, 'Pick up the money.' So I then said, 'If that was YOUR money being wasted day after day, wouldn't you want to pick it up?' They agreed to make the changes being proposed because I had clarified the specific issues that were relevant to the proposal, and intensified the pain.

SWE, the negative elements of your presentation.

Positive elements

Before moving on to the positive elements of your presentation, you could ask how seriously the audience would consider a solution to the problem area you have just highlighted. This will make it easier for you to ask for some action at the end.

The three positive elements are **PRA**:

1 Proposition: the case for your product or service.
2 Reinforcement: reminder of the benefits.
3 Action: what to do next.

Proposition

The time to present the case for your offering is after you have confirmed the existing situation, identified and highlighted the weakness and dwelt on the effect of that weakness. If you have agreement on the argument so far, your audience will be waiting to hear your solution, and your proposition should therefore be presented as that – the solution. It would be a mistake to start giving the history of the company or the origin of the product.

Reinforcement

This is the process of relating the benefits of your offering directly to the needs of your audience. You may have an arsenal of benefits to offer, but the only ones your audience will want to hear about are those that relate to their weaknesses and how you can help them regain or improve their competitive edge. Be brutal in your editing of the list of benefits.

Action

If you have done your job properly, the presentation has caught and held the attention of your audience. You have brought them to the point where they think to themselves, 'This is good. We'd like some of it.' Don't disappoint them.

Tell the audience what they should do next. It's the logical finish to your presentation, and should be specific. How forceful or assumptive you are will depend on the circumstances, but you will be in a strong position if you started your presentation by 'qualifying' them, i.e. by asking your listeners how serious they were about improving their position, in the context of your proposition.

If they have told you that they want such-and-such, and your presentation tells them they can get it from you, how could they refuse?

Essential tip

■ Find out what the audience want and offer it.

Filtering the content

It's always a good idea to start by listing all the points you want to include, and I shall elaborate further in the next chapter, on drafting your presentation. Bearing in mind your purpose, the nature of the audience, and the focus of your message, make a list of everything that is relevant, without editing. Include everything you think of. You can edit later.

(This is similar to, but not the same as, the exercise in the next chapter that is part of the technique for putting your presentation together from scratch.)

I always draw a line down the middle of the page and hand write my list in the left-hand column. When I have finished, and not before, I turn to the right-hand column. In that column I write the benefit either to the audience or to myself. This is the same as asking: 'So what?' about every point I want to make, or answering the question: 'What's in it for me?' (WII FM.)

Although it is fashionable to state that WII FM is the only question to ask, I don't agree. There must also be some perceived benefits for yourself or your company, benefits that the audience can see and hear, because no one believes that a business presentation can be totally altruistic. Customers like to know how you will benefit from the relationship or transaction, and they will want you to be open about it.

For example, you should consider how to emphasise your brand values, and those elements that distinguish your offering from others in the market place. What can you offer that they cannot get from someone else? Can you build in after-sales service that provides a benefit to the customer as well as one for yourself, such as protection of the account? Have you made claims about your offering or your company? If so, your presentation should provide evidence for those claims.

Talk about what you know

On Radio 4 one day I heard an interview with a Buddhist monk. He was talking about a difficult time he was having when preparing for an important sermon he had to give. He wanted to find something original to say, and he had a pile of books on his desk. As he skim read those books, looking for some inspiration, he became increasingly frustrated. He could find nothing.

Finally, he went and spoke to the abbot, explaining his difficulty. The abbot told him, *'You are looking in the wrong place. You are looking in the words and the ideas of others, when you should be looking within yourself. Look within to discover what you want to say.'*

The same advice applies to preparing a business presentation. Look within yourself to realise what you want to say about your subject.

Consider this. If you are making the presentation, it must be because you know something about the subject. What is it? It must be that you have a point of view on the subject. What is that? It may be that you know something about the subject that others do not know, or that your 'take' on it is different from those around you. What is that difference or that 'take'? That's what people will want to hear.

Essential tip

■ Look within for the right message to put across.

Problem solving

One of the more appealing aspects of your proposition may be that it solves a problem that faces your customers and prospective customers. You should therefore set out a problem-solving hierarchy to demonstrate this. You should do this anyway, not just for a specific presentation, because it will help to focus your mind on the specialness of your product or service, and where it can make a difference in the problem-solving process.

It could work something like this:

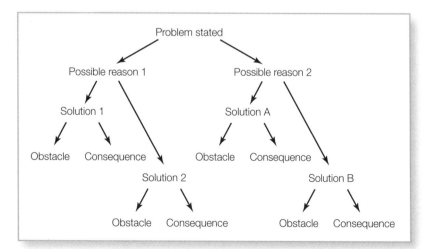

You could then show where your product or service fits in and how it can affect the outcome. That would be another route to deciding what to say in your presentation. The direction that takes should be:

Problem ➔ Pain ➔ Solution

Brand proposition

Every business presentation must reinforce the brand. The brand is the way your company is perceived in the market place. If your presentation is about yourself, if you are a consultant, for example, then it's about how you are known. The brand is about your values, your place in the market, the quality of what you offer, and the instinctive reaction when you or your company are mentioned. Everyone is a brand. It's just that some are better known than others.

■ What's your favourite car, washing machine, soft drink? What is it you like about them, and where do they stand (in your mind) against the alternatives? Now think about famous politicians, footballers, business leaders. What do you think about them? You probably have a clear impression of them, and opinions about them, even if you have never actually met them. That's *branding*.

In simple terms, reinforcing your brand means telling your audience more than what you can do. It's saying, 'We can do this for you *because* . . .' That 'because' is your brand, it's what distinguishes you from alternatives, and your product or solution must always be related to that.

In these competitive times, it is increasingly important for us all to distinguish ourselves from our competitors. To help you focus your mind on your own brand values, here's a *six-point plan*:

1 **Know what you do – for others.** Yes, it is important to know what you do, but don't focus inwards. There is a bit more about this under point 6, but your starting position is to consider what your market wants, and how you can be of commercial benefit to others.

2 **Identify a pain that you can remove.** Is there a weakness that your listeners encounter, to which you have the solution? Focus your 'offering' on magnifying the pain and then showing how you can remove it. You then become the long-awaited Solution.

3 **Do something right.** Apart from removing a pain, there could be something positive that you could do, something that adds to the collective good, something that no one else has thought of doing.

4   Mix with the right people. We all need reinforcement, and we get that from like-minded people, whose own thinking reassures us that we are on the right track. If you network, be selective and don't commit to regular meetings that lead nowhere. If you don't feel uplifted after spending time with certain people, and if they don't understand the things you say, it may be time to move on. Remember, too, that we are judged by the company we keep.

5   Drop the toxic folk. Some people are just plain bad for you. Maybe they are chronically negative, maybe they don't respond well to your enthusiasms, maybe they drag you down in other ways. Leave them to the professional therapists and move along. Don't let them infect your mind or use up your energy.

6   Project your one defining benefit. What's the ONE thing that defines you and distinguishes you from the following pack? Spend time finding out. Get feedback from those you trust. Challenge your first thoughts about it. Then make it the core of your business offerings and everything you say about yourself.

Essential tip

■ Decide what distinguishes you from others and promote that every time.

Is there a take-away?

In deciding what to say and how to put it across, you will want to ensure that your audience leaves with two well-defined things. They are:

1 Your core message, which is the essence of your presentation, expressed in a single sentence.

2 Your brand, which is the set of values that characterise your offering and make it recognisable.

That way they will remember what you can do for them, and why they should get it from you.

You may decide to prepare a *handout*, something for them to take away, so that they remember the essentials. If so, do ensure that the handout clearly states your core message as well as your brand values.

However, it is not a good idea to give them the handout in advance. Some presenters circulate copies of their presentations in advance, to allow the audience to read the slides during the presentation. In the medical profession, they would say that was 'contra-indicated'. In other words, don't do it! I'll explain further in Chapter 7, Using visual aids.

Let's move on now to drafting your presentation.

Summary

- Connect with the audience and what they want to hear
- Talk about what you know
- What will make the audience believe they have problems that you could solve?
- For every fact you want to include, ask: 'So what?'
- Project your brand values to distinguish yourself from alternatives

part

Doing it

5

Drafting your presentation

In this chapter

The hardest part of any presentation is probably the blank sheet of paper. You can sit and stare at it for hours, wondering where to start or writing a few words and deleting them again and again. It's very like what happens if you are asked to make a speech, with little or no warning. You think, 'What shall I talk about?' It's as though you have forgotten all the things you know. I've known even experienced speakers doubt their ability to say anything interesting when faced with the pressure of a deadline.

So where to start, and is there a simple formula that will get you out of trouble? There is, and this chapter will give it to you. However, a word of caution: never follow a formula slavishly, or your presentations will all sound the same and even you will become bored by them.

In the previous chapter I suggested a couple of ways in which you could determine what to say. You should do that no matter which approach you use to construct your presentation. Those techniques, the 'So what?' column and the **problem-solving tree**, for example, will help to clarify your thinking, and should not be confused with what is recommended here.

This chapter is about one particular technique that will simplify the process. I call it *blank page to first draft in 15 minutes*. I developed it specifically for those in a hurry, but it can be used any time to produce a first draft, which you can then refine if you have the time. If you are short of time, just follow this chapter rather than the previous one.

Getting caught out

Here's a typical scenario: you decide to give a presentation about your company's offering, or about some leadership message that belongs to you. You visualise yourself standing before an audience that applauds you and congratulates you on a great message. You feel so good about it that you imagine the presentation is all prepared and in your head.

Time passes and you suddenly realise the presentation is scheduled for the next day and you must give it some thought. So you sit down to sketch out your ideas and plan a few PowerPoint slides. By now the applause is a distant memory as you struggle to put your ideas into some coherent shape, and your enthusiasm for the project runs into the buffers.

The planning phase gets in the way of your purpose and it feels like a chore. On top of that, you realise that your reputation will take a knock if you deliver a so-so performance, and you wish you hadn't taken on the commitment in the first place. Perhaps you also wish you could go from idea to applause in a single bound, without the tough discipline in between.

'Say a few words'

Now consider a different scenario: at a business gathering when you are busy raising your profile and making a good impression on everyone present, your boss or your host turns to you and utters that dreadful invitation, 'Say a few words.'

Your mind races as you search your memory for a joke to get you started. You start to speak and somehow the words flow, but you can tell that no one can make sense of what you are saying, and you are not sure about it either. There's polite applause when you stop, but you wish you knew how to cope with impromptu speaking. You wonder if there is a simple formula. Yes there is, and it's coming up soon.

Something similar could happen when you have a presentation to deliver, but feel you don't have the time to prepare properly. And so you make what could be the worst possible decision: you decide to busk it. You'll make it up as you go along.

Here's the good news. The technique I'm going to show you in this chapter will take you from that terrifying blank page to a usable first draft in a short while, once you get the hang of it. I developed it as a result of two quite different speeches.

Essential tip

■ Master this technique and you'll never be caught out.

Tale of two talks

There was a fellow who had a reputation as a good speaker. Let's call him Henry. He had a background in debating at school and university, his command of the English language was exceptional, and he knew a lot. He was the obvious choice to be best man at a close friend's wedding. Unfortunately, Henry didn't take the assignment seriously, and he neglected to prepare a proper best man's speech.

If you have ever been best man you will know that it is not a role to be taken lightly, but Henry was relaxed about it until the day of the wedding. That's when he asked me what to include in his speech. I rattled off some of the essential elements but Henry wasn't listening. He just wanted to swallow a magic pill and be brilliant.

When he got to his feet, Henry actually announced that he was going to open with a joke. Sadly, the joke was feeble and it not only failed to raise so much as a giggle, but it actually offended some of the more conservative guests, including the vicar.

In show business terms, Henry 'died' on his feet. It was embarrassing and it deeply disappointed the groom and those who had high expectations of Henry's speech. It took years to live down, and Henry was never again asked to be best man.

On a different occasion I took LBC's (London's Biggest Conversation) celebrated psycho-therapist, Sylvia Milton, to a meeting of my Toastmasters club. The chairman seized my arm and said, 'Thank heaven you've come. One of our speakers has dropped out. Could you make a speech this evening?'

I agreed, and retired to a corner to scribble some notes, and produced a speech outline. Although almost impromptu, it won the award for Best Speech of the evening, and apart from the chairman and Sylvia, no one realised it had not been long in the preparation. Later Sylvia said, 'Why don't you work out how you did that and put it in a book?'

This chapter is a distillation of that process.

The difference between my speech and Henry's was that I followed a formula. Henry didn't. Like so many others, he thought he could charm his way through the assignment. It seldom works. A presentation is much more than amplified conversation. You need to consider your own needs as the presenter, and also the needs of your audience.

> *There has to be a match between what you intend to say and what is received and understood by your listeners. And that comes through proper preparation.*

Now here's the health warning about the formula: do not rely on it all the time. It will get you out of a jam, it will get you started even when there is no tight deadline, but it is not the complete answer to preparation. Chapter 4 is about deciding on your content, while this chapter is about structure. But before we work on that, let's consider what enables a presentation to get results.

What to talk about

Let's think about your content and why you are making the presentation. Usually you would be making a presentation because you are expert in the topic or some aspect of it, or because you have some other authority, by virtue of your position (the MD often gets drafted in to make presentations even if s/he are not the real expert).

People don't want to hear the sort of facts they can get from a book, the internet, or someone else. They want to hear *your* take on the subject. They would prefer to hear the world's leading authority on something, and you *are* the world's leading authority on something – on your own point of view. So what's your angle on the subject? Tell them what you think about it and what you want *them* to think about it.

It is important for you to know *what* you know and it is wise to know *that* you know. Suppose you got a phone call this very minute inviting you to appear on television, sitting on a panel of experts on your subject. Would you accept? If not, why not?

It is not necessary to be a 'Mastermind' on every aspect of your subject, able to answer any question that may be thrown at you. But it *is* important to know what you think about it and why. A character in a P.G. Wodehouse novel used to say, 'You've got to have a Nangle.' So what's your Nangle on the subject? See if you can write that down in a single sentence which proclaims you as an expert. Like this:

[Your name] is an authority on [subject] because _____

Essential tip

■ Find your own 'angle' on your topic.

Write your headline

Imagine you are drafting an advertisement or poster for your presentation, and write the headline. Make it tempting, appealing or intriguing. What's the benefit of attending? What can you offer or promise?

Why should you write your presentation title as a headline? For three main reasons:

1 To focus your own mind on the main message.

2 To attract the right people to attend.

3 To lift your audience out of a passive state and into a semi-active state of anticipation.

I've said it before and I'll go on saying it throughout this book, that a presentation's purpose is to bring about change, and that requires persuasion. The first step in persuasion is to engage the attention of your target audience, and a powerful headline-type of title will get you started.

Titles that do not inspire

Here are some seminar titles I found on the internet. You can see for yourself that they could be improved.

- Structured Trade and Export Finance in Russia and the CIS Conference featuring Project and Infrastructure Finance in Russia and the CIS 2011
- 3rd Annual Distressed Investing and Financial Restructuring Australia
- 13th Annual Structured Trade and Export Finance in the Americas Conference
- Corporate Bond and High Yield Market Forum – Indonesia
- Packaging Materials for Packaging Professionals
- Sealing Technology for Packaging Processes
- The Design and Analysis of Fasteners and Bolted Joints
- Life History Optimisation and Seal Breeding Strategies
- Developments in Children's Services Finance
- Communications Based Train Control Seminar

How many of those titles would inspire you to attend? Now here are some *headlines that tick most of the boxes*:

■ They Laughed When I Sat Down at the Piano But When I Started to Play

■ How To Win Friends and Influence People

■ Do You Make these Mistakes in English?

■ Will You Discover the Benefits of AA Membership Only By Accident?

■ How To Be Assertive Without Giving Offence

■ Everything You Need to Know about Direct Mail and Database Marketing

Headlines like these say, 'Stop! Listen, this is for *you!*' Just be careful not to write anything that's too clever-clever or tricksy. Your headline or title must be easily understood, and clearly offer the main benefit.

Essential tip

■ Writing a headline for your presentation will focus its appeal.

Blank page to first draft

This is the shortcut I was talking about. It's the technique for producing a workable outline in a short space of time. It also works as the first step when you have the time to revise and refine your presentation.

You will need three blank sheets of A4 paper

Sheet 1 will carry your outline.

Sheet 2 will be for brainstorming your content.

Sheet 3 will be for your speaking notes.

Sheet 1: Outline

Whenever I ask people how many parts there are in a presentation, they say three: beginning, middle and end. So let's accept that, but rename them A, B, C:

A = Approach.

B = Body.

C = Conclusion.

On Sheet 1 draw two horizontal lines to divide the page into three areas, and label each like this:

A. Approach
B. Body
C. Conclusion

Next, fill in the sub-divisions like this:

A. Approach – Hook – Map
B. Body 1st stream 2nd stream 3rd stream
C. Conclusion – Summary – Action

What you have now is the framework for your presentation, where you can place the relevant ideas from your brainstormed list. I'll come to that in a moment, but let's take a sideways step and talk about the middle section – the **Body**. This is the meat of your presentation and you will see that I have suggested you use a three-part structure, or three streams of argument.

It doesn't matter how many ideas you want to put across, you should group them into three. Common three-parters are:

Past/Present/Future

Past: this is how things used to be.

Present: this is the current situation.

Future: this is what could develop.

Problem/Cause/Solution

Problem: let's define what is wrong.

Cause: how did it come about?

Solution: here are my proposals.

Of course, you could devise your own three-parter, three streams of argument that enable you to build your case towards the proposal you wish to make. You could use chronology, which is a sequence in time, but be careful to accentuate the change that occurred in each period, otherwise it could become 'linear' and therefore boring. Write the chosen themes alongside the three streams, e.g. 1st stream: Problem; 2nd stream: Cause; 3rd stream: Solution.

I have called the first section of the presentation **Approach**, because it contains two vital elements of the beginning: your approach to the start, and your approach to the topic.

Hook: The first thing you need to do is to grab the attention of your audience, just as you would when writing the headline of an advertisement. You need to say or do something remarkable or unexpected, and that's called a hook.

Map: Next, you should let your audience know what you will be talking about. Give them an outline or agenda so that they can follow you, and remember to keep referring to that map as you go through your presentation.

Turning to the final section, **Conclusion**, you will see that I have indicated two elements, **Summary** and **Action**.

Obviously it is always desirable to summarise at the end, to help people remember what you have said, but many presentations omit that all-important call to action. After all, if you are making a proposal, or even if you are just introducing new thinking, you want your audience to do something about it, don't you? Never assume that people will work out for themselves what to do next. They need to be told.

Sheet 2: Brainstorming

At the top of the sheet write your **core message**. That's the single sentence that expresses the idea that you want your audience to carry away with them and remember. Think of it like this: imagine that someone arrives when you have finished your presentation and meets people leaving. They will ask one of them, 'What was that presentation about?' and will typically receive a one-sentence summary.

What would you like that sentence to be? To ensure that it is the *right* sentence, you must write it down first. That's what you put at the top of the page. It is possibly the single most important sentence you will write in your presentation. It provides your focus, it is what you hope and expect your audience to remember, and it sits at the heart of everything you say during the presentation.

Do not put everything you know into the presentation. Applying the Pareto principle, you can reckon that 80 per cent of the impact will come from just 20 per cent of your content, so be prepared to focus on the 20 per cent that really matters, and set aside the rest.

Here's the brainstorming process

Step 1

When you have written the core message at the top, draw a line down the middle, so that you have two columns (see example below). The reason for this is simply to enable you to get all your ideas down on one page.

Step 2

Write down every idea that comes to mind about the core message that you have written at the top of the page, following three rules:

1 No editing. Even wild ideas are acceptable, because they could prompt you to think of other ideas.

2 No full sentences, only trigger words. The two-column layout is designed to discourage long sentences.

3 Number them all.

When you think you have 'dried up', don't quit. Look again at the early ideas and ask questions like, 'How? Where? When?' You could end up with some really good ideas to put on your list. Often the last few are the best.

Brainstorming works by word association, which is why it is essential not to edit. Write down every idea, using trigger words, not sentences. You can delete or ignore some ideas later. At this stage, do not set any limits on the number of ideas you list, or their importance. When eventually you run dry, you will have more than enough material to fill several presentations, so now you have to choose which items to include in this one.

Suppose you were going to give a presentation on networking. Your brainstormed list might look like this:

How to get more out of networking meetings

Core message: Get more referrals by shifting focus from taking to giving

1	Givers and takers	8	Expectations
2	Lunch bill story	9	Providing value to others
3	Disillusioned	10	Not for soliciting work
4	Meetings	11	Network selling
5	Far East way	12	Build good practice
6	Breakfast clubs	13	How others can help you
7	Visibility	14	Good records/database

15 Etiquette at meetings	22 Referrals
16 Questions to ask	23 Checklist
17 Business cards/breadcrumbs	24 Social networks
18 Sharing/enhancing	25 Business networking
19 Post-meeting analysis	26 Online networking
20 Numbers game?	27 Ten-point plan
21 Relationship selling	28 Impatient for results

Step 3: Selecting from your list

The next step is to decide where to place your ideas within the framework above. Place your brainstormed list alongside the framework and write in the numbers from your list.

For example, if you want to use item 2 from your list as your opener, write 2 alongside 'Hook' on the framework. Do that for each section on your framework. (This is why I asked you to number all the ideas on your brainstormed list.)

Let's suppose you want to use Problem/Cause/Solution as your structure. Your framework might look like this:

A. Approach
– Hook 2
– Map
B. Body
1st stream: Problem 3, 8, 22, 28
2nd stream: Cause 6, 7, 25, 20,
3rd stream: Solution 9, 13, 16, 27
C. Conclusion
– Summary
– Action 23

Sheet 3: Speaking notes

Nearly done. The framework above indicates all you want to say on the subject. All that remains is to convert it to speaker's notes. Write notes, just as you would if you were writing on 5×3 cards, with head-line and bullet points for each topic. It could look like this:

Title: Get more from networking meetings

Hook: Lunch story
- arguing over who pays
- takers eat better

Map:
- how to overcome disillusion
- understand why networking can fail
- strategy for more referrals

Problem:
- fed up with networking
- expectations not met
- not getting referrals
- no results after two to three years

Cause:
- breakfast clubs like selling insurance
- must maintain visibility
- in business, takes time to build trust
- not numbers game

Solution:
- focus on value to others
- make clear how others can help you
- ask questions: the right ones
- develop ten-point plan

Summary:
- poor ROI from networking
- probably doing it wrong
- sharp focus on what you offer
- long-term strategy

Action: apply checklist

This is, of course, a short outline, just to illustrate the process. These notes will fit easily on six 5×3 cards, and if you know your subject, if you are already expert in networking, these notes will be enough to acts as prompts as you make your presentation. In fact, they could almost be your slides.

It's worth making yourself familiar with this process, so that you can go from start to finish in about 15 to 20 minutes. It will get you out of trouble if you are ever up against a tight deadline.

Using stories

Now that you have decided on the ideas and information you want to include, you need to add one vital element that will distinguish your presentation from all others – your stories. Facts and figures, events, opinions . . . they are hard to take in and remember, especially if they are overdone. So use stories.

People love stories. They are easier to remember and they illustrate the points you want to make. But tell your own stories, from your own life experiences, and they will remain your own, and make your presentation distinctive. If you must tell someone else's story, always say where you got it.

My friend Paul Joslin attended a speech in the Midlands in which the speaker told a sob story that Paul had heard before in America. When he later tackled the speaker about it, the man said, 'I doubt anyone in that audience has been to America, so it doesn't matter.' Imagine what that did for the speaker's credibility, because Paul told me the tale, and he must also have told many others who, in turn, have told their own contacts.

In this book I have told my own stories, wherever possible, because I know them to be true, and the lessons I draw from them signify my own understanding of those experiences. In Chapter 4 I mentioned Patricia Fripp and the way she constantly mines her own life experiences for material for her speeches. You could do the same. Look at your brainstormed list and think of examples or stories you could use to illustrate some of the points, and add them to the list.

Essential tip

■ Make a point, tell a story – tell a story, make a point.

Now pause for a moment and think about how people receive and remember what you say.

Let us suppose your presentation is planned to run for 30 minutes. If you spoke without interruption, you might average 150 words a minute. That gives you 4,500 words. No one can remember all those words, so the typical listener will follow a pattern, listening first for your general thesis, and then for the structure of your argument, connecting it to whatever they already know or believe about the topic.

The listener will attempt to make a summary and make a decision about what they can use. For that process the listener needs to know:

- What's your point?
- Why is that valid?
- Does the argument hold together?
- Is the evidence sound?

To test your arguments and narrow them down to something usable, there is a three-word phrase that can be used as a question. That phrase is: *Which means that . . .*

Try it out for yourself. Make a claim that you want to include in your presentation, and ask yourself, 'Which means that . . . ?' When you have answered the question, ask the question again, about your answer, and go on asking the question until you are absolutely certain you have said everything you can.

We can move on to consider how to make your presentation persuasive.

Summary

- Write your presentation title as the headline for an ad for your presentation
- Get three blank sheets of A4 paper
- Sheet 1: your presentation outline in three parts: Approach/ Body/Conclusion
- Decide on a three-part structure for the body, e.g., Problem/ Cause/Solution
- Sheet 2: brainstorming for content. Write core message at the top
- Make two columns and write down every idea you can think of relating to the core message. Trigger words only, no editing, and number them all
- Place Sheet 2 alongside Sheet 1 and place selected topics, by their numbers, on Sheet 1, ignoring the rest
- Sheet 3: speaking notes. Write notes as you would on 5x3 cards, with headline and bullet points for each topic
- Select a strong point to use as the hook (attention-getter)
- For the body, follow your chosen three-part structure, to build up interest to the point of desire
- Finish with a clear call to action. Tell the audience what to do next

6

Being persuasive

In this chapter

There are two kinds of business presentations: good ones and bad ones. The bad ones far outnumber the good ones, for a variety of reasons, but mainly because they are not persuasive. Think back to the last business presentation you attended. Did you remain engaged all the way through? Why not?

Were there times when you lost the will to live? Did you pray for it (or your life) to end? And did you work out why?

Chances are you would struggle to remember a presentation that really turned you on. Now pause and do a role reversal. Picture yourself delivering a presentation to people who feel the way you have just remembered feeling about someone else's presentation.

Doh!

I bet you'd rather not be guilty, ever again, of boring your audience. Well, it need never happen again, if you follow the simple guidelines given here.

The purpose of a presentation

Let's start by reminding ourselves of what the purpose of a presentation should be, and then I'll outline the three major considerations that determine the effectiveness of any presentation or speech, and then take you through the seven essentials of persuasion. The rules are broadly the same for both presentations and speeches, except that presentations are usually corporate messages, while speeches are messages that belong to the speaker.

The most important point to remember is that a presentation's main purpose must be to bring about some change. It is not to inform or convey information. It is to make change. If no change is intended, what's the point?

So that's your focus.

It may explain why so many presentations seem pointless and boring. They don't try to bring about change. And they don't follow the rules of persuasion. In fact:

- they are unstructured;
- they are too full of facts;
- the presenter's delivery is poor.

A copywriter's approach to persuasion

I'll deal with each of those weaknesses in a moment, but first let me give you an insight into how some professional copywriters go about the business of writing powerfully persuasive words that induce total strangers to part with their money.

They hold conversations in their heads. When they have a sales message to write, they might spend ages just thinking about a typical person who will read that message. One person. Not a roomful of people. Let me tell you what I used to do, when I was a full-time copywriter. A leading US copywriter told me he does the same sort of thing.

Once I knew all I needed to know about the offer, I would walk around and have an imaginary conversation with the person on the receiving end, trying out different approaches until I came up with one that could make both of us get excited. Then, and only then, would I sit down to write the words that would express the offer and carry it through, quite naturally, to the call for action.

What really made it work was that I was thinking about how the offer worked for the other person, and what would make them excited. I realised that I could never excite someone by writing about what the product is. I would write about what it does . . . for the customer. Too many presentations describe what the product or service is, when they should be talking about what it does.

Essential tip

- Don't describe what your offering is. Describe what it does . . . for customers.

You see, a persuasive presentation (which is just a sales letter in a different form) has to do a number of things:

- it must guide the other person towards a new way of thinking about the topic;
- it must get them involved in the story and its outcome;
- it must influence the way they make decisions;
- it must provide sound reasons to justify the new decision;
- it must make the call for action a natural and logical destination.

Let me now return to the three weaknesses in business presentations.

The importance of structure

An unstructured presentation lacks focus, it wanders all over the subject, and it makes you want to ask, 'Just what's the point you are trying to make?'

Unstructured presentations are hard to follow. The arguments come at you thick and fast like a hail of buckshot, and you are expected to decide how they connect to one another, even while you are still listening to the rest of the presentation.

Clearly, then, we must ensure that our own presentations are well structured, and structure will be one of the main building blocks when we come to discuss the techniques for putting presentations together.

Filter the facts

The second reason why presentations fail is that they are too full of facts. Now I know that some people would say, 'But surely the purpose of a presentation is to put across facts?' That's a popular impression, but it's a false one. If your intention is to communicate facts alone, then why not email it to your audience? That would be far more efficient.

If you are going to make a presentation in person, your audience will want you to interpret the information. They'll want you to say, 'This is what I think about the information. This is what I want you to think about it. And this is what you can do about it.'

Why delivery is important

Does it matter how well you present? Well, think about TV broadcasters, such as the people who read the news. Do you like some of them and not others? Think about what happens when the news channel finishes the national and international news, and switches to local news. Often the newscaster will be less experienced, and not as good as the people reading the main news.

You notice the difference, don't you?

There is always a difference between one person's delivery and another's. It directly affects the way we receive what is being said. Exactly the same thing happens during a presentation.

You have to be a 'main news broadcaster' when you present, and that comes from conviction, confidence and practice. Don't imagine you can shelter behind your content. The way you deliver your presentation will determine how the message is received and accepted.

Delivery is much more than the way you pronounce the words or speak your script. It's about the way you put yourself into the message and make it personal. It's about the connection that you can make with a live audience. It's about the way you can touch their feelings, change their thinking, inspire them to a new way of doing things.

Essential tip

■ Connect with your audience by speaking with conviction.

What outcome?

So the questions you need to ask yourself at the very beginning are, 'What's the message I want people to carry away?' and 'What outcome do I want? What change do I want to bring about?'

The answers to those questions will go a long way towards helping you to make a presentation that people will want to hear.

But where should you start? What makes a presentation succeed? What are the key ingredients?

A good place to start is to think about how your proposition will benefit the other person.

What do people expect from you?

When people sit in front of you to receive your presentation, they do so for one reason: they believe you have something of benefit to offer. They do not want to hear about you, your factory, or how many years you have been in business. They want to know what you can do for them.

Now, here's the really interesting bit: they usually do not know they need you. If they did, they'd come looking for you. So you have the opportunity to open their eyes to a problem they may not know they had.

Your opening should therefore be to establish that need. It's the first step in persuasion.

Here's a simple way to arrive at a good opening. Answer these questions:

- What can I offer that they haven't got?
- What are they losing by not having it?
- What will they gain by having it from me?

Persuasion works. In the past (and still in the remaining autocratic nations) the public at large did as they were told. It was coercion. Persuasion is the non-violent alternative and, in the long term, much more effective.

Social interest theorists define persuasion as a form of social influence. Marketers define it as the technique for changing behaviour. Either way, it is about people embracing new thinking and new ways because they come to accept that they will benefit.

Persuasion is so powerful that as long ago as 1956 Vance Packard wrote his seminal work, *The Hidden Persuaders*, in which he warned of the dangers of manipulation. Those concerns were echoed by Dr J.A.C. Brown in his 1963 book, *Techniques of Persuasion*. Since those days it has been recognised that persuasion and manipulation are not the same thing.

There have been many studies of the phenomenon of persuasion, as it applies in politics, negotiation, advertising and elsewhere, and all agree on one thing: persuasion works. It is about changing attitudes, beliefs and behaviours.

The major difference between persuasion and manipulation is this: persuasion leads people to change for their own good, while manipulation brings about change for the benefit of the person who wants the change to happen. Persuasion is for the benefit of others. Manipulation would be for your own benefit.

Essential tip

■ Focus on how the audience will benefit, rather than how you will.

Persuasion can be carried out through words, action or a combination of both. It works best when there is benefit on both sides. This is explained by Robert B. Cialdini's Law of Reciprocity – the tendency of people to return favours. Do something for others and they will want to do something for you.

An example of this happened to me in Scandinavia.

An unexpected challenge

It could have been a disaster. I had done some work for a large high-tech organisation in Oslo, and it had been well received. They invited me back to deliver a course on advanced presentation skills, and I was presented with eight delegates.

Half an hour into the course I sensed that all was not well, so I stopped and asked them what was on their minds. They told me that they were all engineers, and could not see the benefit of improving their communication skills, as the work they did was results-based and spoke for itself. Apparently the original delegates had dropped out because of internal problems and politics, and the engineers had been dragooned into replacing them. They did not really want to be there.

I had to speak their language, or the day was dead in the water, and I was a long way from home!

I asked if they would like to see a live example of the technique for putting together a credible talk in a hurry. They said 'Yes'. When I offered to deliver a talk on a subject of *their* choice, their eyes lit up at the challenge. I asked them questions about it, placing their answers on a mind map on the white board behind me. They were intrigued, never having seen a mind map before.

Persuaded by the demonstration

My questioning enabled me to identify a core message that I could focus on, and I delivered a short talk on that, using the mind map as my prompt. The eight delegates were sufficiently engaged to stay for the rest of the day, and they all agreed they got value from it.

What was it that rescued the day? Two things:

- First, I showed them a technique (mind mapping) that they had not seen before, but which appealed to their way of thinking.
- Second, I demonstrated the application of what I was proposing to teach them. In a sense, I spoke their language (providing proof), and that did the job of persuading them to listen.

For a presentation to succeed, it must persuade, it must bring about some change. Persuasion is about overcoming preconceptions, it's about inducing people to change their positions or their attitudes, and then doing something about it.

It's not enough to convince them. Convincing is only about changing their thinking. New thinking could be academic, and it is not enough. Persuasion is about getting people to take action, to do something new, or to do things differently from before.

Essential tip

- It's not enough to change the listeners' thinking. You must get them to act.

Seven essentials of persuasion

The Scandinavian episode met the **seven essentials of persuasion**:

1 **Listening.** I first asked them what they wanted and paid attention to their concerns.

2 **Relevance.** The subject matter of my talk was directly relevant to them because they had supplied the content themselves. What made it especially significant was the fact that, even though I am not an engineer and do not know their subject beyond what they told me, I was able to deliver an acceptable talk on it, using my method.

3 **Alternative.** You must offer an alternative to the current practice. My mind map was different from, and more efficient than, their

current mode of preparing a presentation or report. In addition, from their answers to my questions I was able to arrive at a new angle on the topic, one that had not occurred to them. That made it new. And it made it my own.

4 **Meeting expectations.** Apart from the questions on factual content, I had asked them what they wanted, and what would interest them. They were involved throughout and their expectations were exceeded.

5 **Trust.** As engineers, they believed in evidence. They wanted proof that my course and my techniques would work and would benefit them.

6 **AIDA.** I followed the classic sequence of persuasion, which is detailed below.

7 **Commitment.** You need to ask for the commitment. At the start I asked them what they would like me to do. At the end, having demonstrated the proof, I asked, 'Shall we resume the course?' and they agreed.

Let me expand a bit on the seven essentials.

Listening

I am grateful to Jack Pachuta of Management Strategies Inc. for a surprising statistic. He says that the communication skill we use most is the one that is taught the least. In any day, he says, an individual communicates through the spoken word 30 per cent of the time, in writing 9 per cent of the time, they read 16 per cent of the time, and *listen an amazing 45 per cent of the time!*

Yet, how many people have been taught *to be effective listeners*?

You can encourage or block the other person, simply through your body language. When you lean forward, maintain good eye contact, nod repeatedly and remain relaxed, the other person becomes more articulate. When you are tense and itching to say your piece, or when you are switched off, the other struggles to express themselves.

When you are asking about your prospect's needs, listen carefully, nod or mutter 'Hmm', and repeat a summary of what they have said, using their own words as much as possible, to indicate that you have understood.

Ask questions for clarification if necessary, then ask what they want you to do about the point they have made. If necessary, re-state the point briefly in your own words, and link it to your own message.

Note-taking is helpful in listening well, especially during the Q&A session. Always welcome questions when you present, because they give you the chance to re-state your message when you answer.

Essential tip

▓ Develop the skill of constructive listening. It will distinguish you from others.

Relevance

An IT-related company in London wanted help to improve their sales. Their expectation was that it would be enough simply to restructure the sequence of the slides in their standard sales presentation and boost the confidence of the presenters with a morale injection. Many client companies have the same limited expectations.

When asked about the company's offering, the general manager spoke at length and in detail about what it did, and about how it improved on what was there before. But he quickly became bogged down in technical details. In addition, whoever produced their slides clearly did not have a sales background. The slides were full of graphs and tables, and were very text-heavy. Not only that, they proclaimed the company's technical expertise in a very self-centred way.

The presenters themselves had no knowledge of the process of persuasion. They believed it was enough to describe their product in technical terms, and to parade their slide presentation. It was hardly surprising that their conversion rates were dire.

It was not easy to wean them off that approach, because their thinking was based on 'description' rather than 'benefit', i.e. 'This is what the product will do' rather than 'This is what it will do *for you*'. Adding those final two words to their thinking made all the difference.

Alternative

It's a well-known mantra that if you go on doing what you've always done, you'll go on getting what you've always got. How that applies will depend on the kind of audience you are addressing. There are two kinds. One kind accepts that things could be better, and they are more likely to consider your proposition. The other kind is quite content with the way they do things, and will need a more powerful opening to get them to listen to you.

Sometimes it's worth taking a big step back and looking at what they are trying to achieve with their current practices or the products they buy. The alternative may not always be another version of the same thing. For example, what's the alternative to a Mont Blanc pen? It's probably not a Montegrappa or some other pen. The Mont Blanc is frequently bought as an executive gift. So the alternative could be some other executive gift, rather than another brand of pen.

Apply the same thinking to your own offering, and see if you can make it relevant by relating it to your audience's needs rather than by comparing it directly with their current solution.

Meeting expectations

Each nation has its own set of expectations when preparing or receiving a presentation. These are commonly believed to be the expectations of Americans (on the left) and Brits (on the right). I am not recommending these sequences, merely reporting that these are the common expectations in the US and UK.

US	UK
■ Humour	■ Humour
■ Joking	■ Surprising fact(s)
■ Open with a wow	■ A story
■ Modernity	■ 'Nice' product
■ Gimmicks	■ What make it tick
■ Slogans	■ Reasonable price
■ Catchphrases	■ Quality
■ Hard sell	■ Traditional not modern
■ You deserve the best	■ Neat summary at end
■ Go out with a bang	
Attention span: 30 minutes	Attention span: 30–45 minutes

While it is not necessary to follow the sequence shown, it is worth bearing in mind, because you will need to be strong about retaining interest when you stray from the expected path.

Trust

The four pillars of trust are said to be:

1 Reliability.
2 Capability.
3 Honesty.
4 Empathy.

In addition, the audience must *like* you.

Trust is essential within a business, as between employers and employees, and it is increasingly a factor in determining buying behaviour. Clients and customers choose suppliers they trust and quickly jump ship when trust is broken.

Stephen Covey has written a book called *The Speed of Trust*, in which he says that, in these competitive times, speed in business is essential, and nothing affects speed so much as trust. Without it, everything slows down. Trust, he maintains, creates speed more effectively than anything else. It can reduce costs, increase productivity, enhance innovation, improve collaboration and increase value.

In fact, trust creates the climate of willingness for change.

Essential tip

▪ Business is based on trust. For that the audience need to like you.

Let me now turn to the classic sequence of persuasion.

AIDA

Before anyone will accept your proposition, their interest must reach a minimum level, which I call the **buying level**. At the start of your presentation you should assume that your audience's interest is somewhere between zero and the buying level (see the figure below).

Attention (A)

The first step is to gain your prospect's involved attention. You need to say or do something that signals: 'Stop! This is for you!' It must therefore be relevant to the prospect's business or needs. In a press advertisement, that's the job of the headline. In a presentation, it's the job of the hook.

Interest (I)

Your hook should raise interest sharply, but it's unlikely to reach the buying level in a single bound. You now need to develop the prospect's interest in your proposition by providing more information. If you offer facts and figures, always interpret them as benefits. And every time you make a statement or claim, or throw in a statistic or figure, answer the question: '*So what?*'

No one buys on facts and figures alone – those are only used to justify the decision to buy.

So pile on the benefits until the prospect signals, 'I'd really like to have this!'

That's the point of Desire (D)

This has to be well above the buying level because at some point you will have to reveal the cost, and it's not always a cash cost. There's inertia cost as well, the reluctance to change. As you know, when you reveal the cost, interest is likely to drop. As long as it does not drop below the buying level, you can reassure and reinforce your proposition, sending the arrow back upwards again.

Add up the benefits of your proposition, making the package seem as large as possible, before discussing the cost. Remember that there will always be an inclination to continue as before. To overcome that inertia, you must build up anxieties about the present situation, and then expand on the benefits of change. The pile of benefits that follow change must be considerably larger than the cost of the change you are proposing.

Finally, seek agreement and move naturally to asking for the commitment.

Action (A)

Always be prepared to ask for action. Never leave your audience hanging in the air, having become excited by your proposition. Make it easy for them to take the next step, but tell them what they should do.

Here is a graph indicating the AIDA sequence:

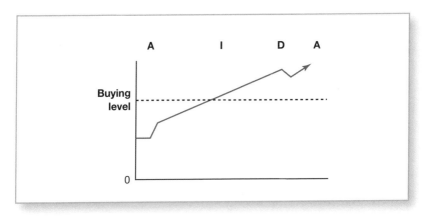

Commitment

The words you use will vary from situation to situation, but the logic is clear: *you have agreed that this situation is unsatisfactory and is costing you this much, here is a solution that will remove those costs and disadvantages, and this is what you should do to make that happen.*

Many presenters are afraid of asking for the commitment at the end of a presentation, but there is no need to hesitate. If the presentation has been properly conducted, with incremental agreement about the improvements that are required, the prospect(s) may actually *expect* to be asked for the commitment.

Next up will be visual aids, the bane of many a presenter's life. No need to fear 'Death by PowerPoint'. There is a better way.

So what should you remember and carry in your head about persuasion?

Summary

- AIDA: Attention, Interest, Desire, Action
- Describe what your offering *does*, not what it *is*
- Develop trust
- You cannot live by fact alone. People buy on emotion and justify with reason
- Lead people to *want* what you are offering

7

Using visual aids

In this chapter

- The three main types of visual aids: those you project/create/show

- Making the most of PowerPoint

- Adding video or audio to your slides

- How to use a flip chart

- Using props

- Avoiding common mistakes

I n this chapter, we'll be looking at the *three main kinds of visual aids*, how to put them together and use them, as well as how many slides to have, and how to avoid common mistakes. Visual aids help you be persuasive, because they engage the sense of sight, as well as sound. When you engage all the senses you involve the listener totally.

There is more to visual aids than PowerPoint, but we can start with slides and how to use them. PowerPoint is a relatively new technology, developed in 1987 as *The Presenter* and upgraded to its present form ten years later, featuring transitions that rescued presentations from their linear structure, one image at a time, as in the use of overhead transparencies.

I had a call from a man who was offering training in some aspect of computer technology. He was happy with his content, but just wanted some help with platform technique – how to present with his slides.

Not wanting to carry his laptop, he arrived with his slides printed on overhead transparencies. As he took them from his briefcase, I said, 'I think you have too many slides.' He looked astonished, and said, 'There's only 40!' For a presentation that was planned to last a little over half an hour, that was clearly too many, so we worked on cutting them down. As a simple rule of thumb, think of an average of two minutes per slide, although it does depend on what's on the slides. If you are showing a succession of photographs, you can probably have more than text slides.

I pointed out that with so many slides, he might as well run a video.

Three months later he rang me and said, 'I need help. I've now got 108 slides.' He was illustrating every point he wanted to make, and would have been better off turning them into a stand-alone slideshow with text overlays and music.

That's what Steve Kayser did when he had to produce a presentation for a budget committee, on what his PR department had achieved during 2009, in the tough economic climate that prevailed then. He was allotted five minutes. But when he had created all the slides he needed to tell the story, he had 118 of them: 118 slides in five minutes. So he put them in a slideshow, with music, and stood silent while it ran. It worked, and the budget committee even approved two additional staff on the team. To see what Steve produced, go here: **http://bit.ly/tuAJX**.

Apple's former marketing chief, Guy Kawasaki, uses what he calls the *10-20-30 rule*:

■ **10 slides** (one for each of 10 main points).
■ **20 minutes** (length of presentation).
■ **30-point text** (minimum size).

In fact, you could consider even larger type, and see how it looks when you try to cram the usual number of words on the slide. According to one study, the average number of words on a presentation slide is 40. In my own experience, that's a serious underestimate, and I'll deal with the effect of that in the next chapter, where I address the matter of how audiences listen.

Guy Kawasaki doesn't believe in long presentations, especially if they are not structured. He once said, '*I discovered that if there's anything worse than a speaker who sucks, it's a speaker who sucks and you have no idea how much longer he or she is going to suck.*' He added that all his presentations are in his 'top ten' format (as above), so that you always know how much longer he is going to be.

Types of visual aids

In broad terms, there are only three types of visual aids in general use with business presentations:

1 **Those you project** (slides, film, video, audio).
2 **Those you create** (flip chart, white board interactivity).
3 **Those you show** (props, objects).

I'll deal with each in turn, but the general point to make about visual aids is this: they are *aids*. Their function is to assist, but not supplant, the presenter. The presenter – you – must always be the presentation. No visual aid is more important or more powerful than a live presenter, and should never be allowed to become the most important part of the presentation.

Inevitably, there will be some who search their minds for *exceptions*, and one example might be when a presentation includes a contribution, on screen, from someone at a remote location. On such an occasion, the screen becomes central to the presentation, but the usual rules apply to the remote presenter. In effect you stand aside and they take centre stage.

Do not get carried away by the possibilities of technology. Your purpose is not to dazzle your audience with the gizmos at your disposal, but rather to inspire in them a sense of wonder at your way of looking at something they may have taken for granted. The magic of your imagination is what makes a presentation compelling, and the visual aids are there merely to help bring your ideas to life.

Good enough is not enough

I once saw a sensational presentation at the National Speakers Association in America, given by a National Geographic photographer called Dewitt Jones. The photographs he showed were breathtaking, but not because of their technical excellence, which any good photographer with the right equipment could match. What set his pictures apart was his perception, his viewpoint, and the patience to wait for the 'great' shot when he already had a 'good' one. It was what he called *'seeing the extraordinary in the ordinary'*.

Show people how to see something differently, and you'll have their attention. If you can arouse the curiosity of your audience, they will hunger for your solution. You can do that with pictures, with words, with sounds. Don't search for perfection, because there is usually more than one right answer. Find your own right answer, and use whatever it takes to bring it to life for your listeners. That's the essence of creativity, and that is what will elevate your presentation high above the PowerPoint platitudes that others present.

Essential tip

■ There's usually more than one right answer. Find your own.

The Greek orators of old used no visual aids. They relied solely on the power of their words and delivery, and many a present-day speaker does the same. President Obama is one such example. But people remember better if the experience is multi-sensory, if they see as well as hear, and even more so if they can feel or touch as well.

Projecting slides or film

PowerPoint is the most common form of slides used for presentations, and is often derided because it is not properly used. 'Death by PowerPoint' is a favourite term used to put down slide-heavy presentations. However, if certain rules are followed, PowerPoint can provide a useful vehicle that's easy to use. PowerPoint has a number of standard slide designs that conform to best practice.

Essential rules for slides

1 **Information slide:** headline and supporting points.

2 **Word maximum:** 5x5 (5 lines, 5 words per line).

3 **Topic slide:** fewer than 5 words.

4 **Idea slide:** picture with caption or limited words.

The function of a slide is simple. It is to provide a visual summary of what is being spoken by the presenter. Apart from graphs and other slides that contain complex information, it should be possible to glance at the slide and know in a moment what it means, and how it summarises what is being said. Like this:

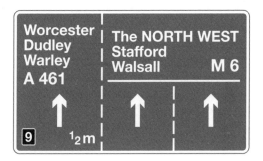

© Crown copyright

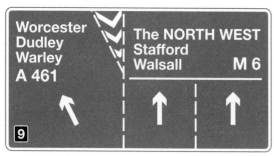

© Crown copyright

When you are driving along the motorway, at perhaps 70 mph, you need to take in the information on the sign very quickly. The two signs above, which you will see half a mile apart, tell you quite clearly what to expect and what to do if your destination is Worcester. Not all road signs are this clear, however, and you may sometimes have to drive more than once around a roundabout to find the right exit!

Here are a couple of slides that achieve their objectives by getting across their messages in an instant.

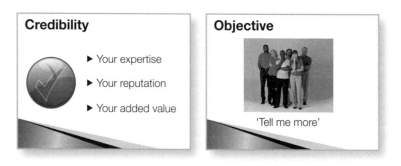

Try this little experiment. Switch on the radio, and tune to a non-music channel, with the volume turned up to normal conversational level, or have someone tell you a story, while you read a passage from a book or magazine. Try to listen and read at the same time, as though you were going to summarise both at the end of one minute. How long were you able to do it? Most people quit after a few seconds, because it just cannot be done. Yet that is what you will be expecting your listeners to do if your slide cannot be taken in instantly.

No full sentences, and use charts sparingly.

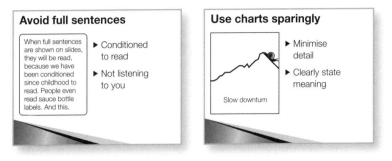

In the left-hand slide above, the small words in the box will be read, simply because we have been conditioned to read everything. Is that an argument for putting even more words on the slide? No, because you want people to listen to you, and not drift away while they read the slide. Mind you, if the text is too small to read, people become irritated and switch off.

Treat your information slides like press ads. Write a *headline* that clearly states what it is about. If you want to use bullet points, that's fine, even though some people hate them. Bullet points should not be more than five words long, and there should not be more than five lines on the slide. Bring up the bullet points one at a time, to help your listeners concentrate on what you are saying. The slide design must conform to the way people are conditioned to read, i.e. left to right, top to bottom (right to left in some countries).

Why five? Because research has shown that people can remember an average of seven units of information, plus or minus two. Therefore, five is within everyone's scope.

Avoid punctuation and capital letters. Email has conditioned us all to regard CAPITALS AS SHOUTING, so they are best avoided. Punctuation causes breaks in the flow, and gets in the way of the message.

Some tips on design:

■ make every slide look different, even if you are using a design template, otherwise your listeners will have difficulty realising that they are not still looking at the previous slide;

■ use the same colour scheme for text;

■ avoid strongly coloured or fancy backgrounds;

■ remember that red signals danger or warning and use it sparingly;

■ if using pictures of people, position them to face into the page. A face turned towards the right should be placed on the left. The viewer's eyes will follow the gaze of the person on the screen, and you don't want them leaving the page;

- follow the natural inclination of the viewer, whose eyes naturally land one quarter of the way down from the top, run left to right, and down to the bottom right-hand corner – good direct response ads follow this pattern;

- use pictures wherever possible, either on their own or with brief captions, and preferably your own pictures, not ClipArt or stock photos;

- use large type, even on graphs, and check that it is all readable from the back of the room;

- use sans serif fonts such as Arial and Verdana, because they are bolder than serif fonts and better suited to large projection. Serif fonts were developed to help the eye to travel along the page of a book, in which type is small.

Transitions

Some people get rather carried away with transitions and animations. Their slides zoom in, explode on the screen, materialise out of a checkerboard, enter from all sides, and generally leave the viewer dizzy. Use transitions and animations sparingly. Your presentation is not a fireworks display, and the transitions should be used to serve a purpose, not just because they are available.

Finally, take a step back from the presentation and ask yourself if they look like a unified whole, like a set of smart, matching luggage, or are they like a tray of mixed cakes, each one different, a kaleidoscope of visually confusing images.

Number of slides

Some people reckon one slide per minute, but I work on an average of two minutes per slide, as I said earlier, but you should switch off the slides altogether from time to time, and re-engage with your audience.

Managing slides

I was at a seminar recently which was conducted with a PowerPoint presentation. After the first half hour or so, someone in the audience disputed a piece of information that was on one of the slides, and a lengthy discussion ensued, drifting away from the message on the slide, which stayed on the screen throughout, acting as a visual distraction. When the presentation resumed, another delegate asked if the presenter could go back to an earlier slide. He scrolled back through all the slides until he found it. Not very professional.

In the comfort break that followed, I approached the presenter and showed him three simple ways to control the mechanics of the presentation:

1 To switch off the slides while a discussion is taking place, press letter B. The screen goes blank and attention reverts to the presenter. B for Black or Blank. Press B again to return to where you were.

2 To switch off the slide but keep a bright screen, if for example the room lights have been dimmed, press letter W (for White). The slide disappears, but the screen remains lit. Press W again to return.

3 To go directly to any slide, have a printout of all the slides, numbered. Type in the number of the slide you want and press Return.

Essential tip

■ Switch off the PowerPoint occasionally, to return attention to you.

One more thing

Always check that your slide presentation can run on any computer, especially if you are loading your presentation onto a CD or memory stick. Some computers have a different version of PowerPoint from your own, and some computers may not have PowerPoint at all, so you should have PowerPoint Viewer on your CD or memory stick as well.

Video or film clips

■ Keep them short: people's attention spans are very limited.

■ Use only professionally produced clips, topped and tailed: we have all been trained by TV and the cinema to expect certain standards, so you should have Title and Presenter at the start and contact details at the end.

■ Have a soundtrack: a silent film is disconcerting.

The biggest danger associated with film or video clips during a presentation is that they tend to interrupt your connection with the audience because:

■ *you* are not doing it;

■ the lights are usually dimmed;

- they are received as 'entertainment';
- the sound is different from you.

Video can be incorporated into PowerPoint presentations, but only if you follow the guidelines. It can be used as self-contained clips at given points of the presentation or as a background. The problem with video backgrounds is speed. Set the video background to a slow speed so that it doesn't jar with the usual slow speed of slide introductions and animations, and use video selectively.

The video itself must be in a format that fits with PowerPoint. It can be QuickTime, Flash, AVI or MPEG, but the format that works best is Windows Media Program (WMV) which, like PowerPoint, was created by Microsoft itself. WMV is also the format used by YouTube.

To play video within a PowerPoint presentation

First create a WMV file for your video clip and place it on your desktop, then open your PowerPoint presentation (not through PowerPoint Viewer) to the slide where you want to insert the clip:

1 Click on Insert (drop down menu).

2 Select Movie.

3 Browse on Desktop and select video clip.

4 Decide if video should start immediately when you reach that slide or when you click it.

The same instructions apply if you want to insert a sound clip rather than a video, for example if you want to add applause or music to a particular slide, but remember to take advice on copyright if you are going to use someone else's original music.

Essential tip

- If adding a video clip, first make it right for YouTube (WMV).

The visual aids you create

The simplest kind of visual aid is the flip chart, on which you might write words or draw a diagram. An alternative would be the white board, but the principles are the same. If you are skilled at it, this visual aid can be very effective, because it is live, it derives from the current discussion, and it grows in line with the conversation.

The seven advantages of a flip chart are:

1 They can be used anywhere.

2 They do not require electricity.

3 They are directly relevant.

4 You can use colour.

5 They are economical.

6 They encourage interaction.

7 You can combine prepared content with new.

Here are the guidelines:

■ Use only if your writing or drawing can be clearly seen and understood from the back of the room. I once saw a presenter using a flip chart when presenting to 300 people. Predictably, it failed.

■ Use broad nib (chisel tip) pens. Always carry your own as most venues provide the wrong kind.

■ Write LARGE (not less than 2in/5cm high).

■ If you are going to draw a picture or diagram, lightly draw it in pencil in advance so that you produce an acceptable image that does not diminish you in the eyes of your audience.

■ If possible, do the same for text you plan to write, using a ruler to keep the lines straight, or draw tramlines in pencil, 2in/5cm apart, to keep your writing straight and consistently large.

■ Place the flip chart so that you do not stand with your back to the audience when writing on it.

■ Practise writing *fast* and legibly. It's not the same as writing on a page laid flat on your desk.

■ DO NOT USE ALL CAPITALS. Upper and lower case is easier to read.

■ Have some Blu-Tack ready in case you want to stick some completed pages up on the wall.

■ If you write on any pages of the flip chart in advance, make sure you leave them covered by at least one blank page, so that they do not show through.

Essential tip

■ Write or draw in pencil on the flip chart in advance.

A different kind of 'created' visual aid is any *interactive participation* by audience members. When you ask people to come forward and do something, they become visual aids, and it is important to manage the process well. Here are some rules:

- Treat every participant with respect.
- Never make a person feel uncomfortable.
- Remember that people in the spotlight can become tense or nervous, and may sometimes behave unpredictably.
- Always ensure that they leave the platform feeling good about themselves, and lead the applause for their contribution.

Use of props

Any object that you use to illustrate a point is a visual aid. It could be a paper aeroplane, an apple, a laptop, anything at all, even a rubber band. Apple's Steve Jobs usually uses some prop, and when he introduced the new, slim Macbook Air, he held up one of those large manila envelopes for inter-office memos, undid the cord and slid out a Macbook, thus demonstrating how thin and light it was.

Props can be used metaphorically or to demonstrate some feature of the product you are describing. When I produced an apple and crossbow during my talk to dentists about rapport, it was a metaphor for trust, not about marksmanship with a bow.

On another occasion, when talking about getting rid of the limiting labels of the past, I said, 'Write them on a sheet of paper and make a paper aeroplane with it, then go to some high place, throw the plane over the edge and let those nasty labels fly away.' As I spoke I produced a paper plane I had prepared in advance and threw it down the aisle. People remember those things and reconstruct your message from them.

On another occasion, I was presenting to advertising people, about direct marketing, and used a small model aeroplane that I could throw and have it return and settle on my outstretched hand. A colleague later told me, 'I bet everyone expected that to fail!' But I had practised repeatedly in advance and knew it would succeed.

Essential tip

- When using props, practise so it can't go wrong.

Essential guide to avoiding common mistakes

Too many slides: work on an average of two minutes a slide, and one minute per slide as a maximum. More than that and you might as well project a video. Are there exceptions to the rule? Of course, but only if you have a section in which several slides are projected in rapid succession, as when you are showing different examples of something.

Hard to read: Use large type (see Kawasaki, p. 89), and always check that EVERY word can be read from the back of the room. Use sans serif fonts, as they are bolder than serif fonts.

Too much information: Always work on the three-second rule. Pick any slide that has a lot of content, and test it on a colleague by showing it for just three seconds and asking them to tell you what they got from it. They don't have to know the detail, as long as they understand the main point of the slide. Remember, too, that if they have to take time to read and understand the detail, they will not be listening to you while they do that. Follow the 5x5 rule as well, which is 5 lines max. and 5 words max. per line.

Slide dependency: It is not essential to have slides. Some of the world's top motivational speakers use no visual aids at all. Don't start your preparations from the slides. Draft the presentation and try delivering it without any, until you feel the need for illustration. That's where to have a slide.

Unbalanced video clip: A film/video clip that is too long can kill the presentation stone dead, so keep it brief. Also, consider how it fits with the rest of the presentation, in terms of rhythm, sound, colour and pace. If you need to turn down the lights, you risk sending some people to sleep.

Illegible writing on flip chart: Use broad nib (chisel tip) pens, write large (2in high letters) and use pencil tramlines to keep your writing straight and consistently large.

Embarrassing participants: Every delegate is your customer and must not be made to feel uncomfortable. If you use delegates to demonstrate something negative, always return them to a positive state before releasing them. And never push them to do something they feel reluctant to do, even if it is something simple. Stage fright can seize anyone, and they should be helped to save face.

Props that fail: Always rehearse the use of props. I saw a very experienced speaker try to do the three-rope trick and fail because he had not practised it enough. He not only failed to make the point of the trick, but he lost some ground with the audience.

Let's now consider how audiences listen, so that you can connect with them more effectively and maintain their interest.

Summary

- Guy Kawasaki's 10-20-30 rule
- Visual aids you project, create or show
- Slide rules
- How to use video clips
- How to use a flip chart well
- The common mistakes to avoid

Connecting with the audience

In this chapter

A s you approach the point of having to stand and deliver, chances are you'll have an attack of the nerves. For some reason, when the spotlight is on us we lose confidence in our ability to deliver an impressive performance. Let's be very clear about this. It is a *performance anxiety*. So let me offer you a seven-point strategy for bringing those nerves under control.

First, identify what your fears really are. Are you afraid that:

- people might laugh;
- you might dry up;
- people will stop listening;
- your voice will break;
- people will be hostile;
- you will not meet expectations;
- you'll make a fool of yourself;
- you will lose face or status?

Seven-point strategy for overcoming nervousness

1 Remind yourself that you know what you are talking about. Why is it you speaking? What do you know that the audience don't know?

2 Thoroughly prepare and trust your material. If you know what you want to talk about, and you have all the facts and ideas ready, you can't be caught out.

3 Keep repeating your core message. It will give you focus. Keep saying to yourself, 'What I want them to hear, understand and accept is . . .'

4 Do not try to deliver the whole encyclopaedia on your subject. Keep it simple. You can deliver an entire presentation on just one point. Or you can make it two or three points. That's enough for any audience.

5 Do some proper deep breathing. Take a deep breath, then breathe out and when you think you have emptied your lungs, blow out six imaginary candles. Then breathe in again. Do this three times just before it's time to speak. It will clear your head.

6 Raise your metabolism. Go somewhere private, throw your arms up in the air and cry, 'YES!!' Do that three times. If you cannot go somewhere private, grip your seat between your legs and pull upwards as hard as you can, and release. Do that three times, just before it's your turn to speak.

7 Finally, recall some incident involving a child or an animal, an incident that makes you smile. That will create an inner smile, which will make you look more approachable, and your audience will respond with friendly smiles.

FEAR stands for:

False

Evidence

Appearing

Real

Be worried if you're NOT nervous. Are you properly tuned in?

- Use the adrenalin. It keeps you on your toes.
- Always make the effort to persuade your audience.
- Never take an audience for granted.
- Tune in to what the audience want and find the energy to provide it.

Your audience consists of people just like you. They have the same doubts and anxieties as you, but above all, they want you to succeed. No one attends a presentation wanting the presenter to fail. So they are on your side and will help you along, if necessary. Incidentally, even the most experienced speakers have nerves and performance anxiety before they speak. Once they get started (as you will probably find yourself) they are fine.

When I was at university, the lecturers would stride back and forth at the front of the room, often with hands behind their backs, sometimes pausing to stare out of the window while still delivering their wisdom for the benefit of anyone who cared to listen. They weren't much bothered either way, and quite a few of the students were equally uninvolved, slipping out of the lecture room soon after registration. As examples of communication, those lectures were poor, and we were left to pick the bones out of their content, supplementing them with our own reading.

How audiences listen

I didn't realise it at the time, but quite a few of those lecturers had developed that disconnected, uncaring style out of performance anxiety. If business presentations were delivered in the same way, they would be a total waste of time. It is important to connect with the audience, and recognise how they listen and understand. Remember that people listen:

- for a confirmation of what they already know;
- to obtain new information;
- to learn and understand;
- to be entertained.

As I explained in Chapter 4, your audience first want to know that you are on the same page as they are, that you have *knowledge in common*, that your starting position is their existing fund of knowledge, and that you will extend that knowledge. Unfortunately, even when the subject is of interest to them, they will remember only 25 to 50 per cent (or even less) of what they hear, within a few days of your presentation.

Essential tip

- You and your audience have common knowledge. That's your launch pad.

Every person has their own way of listening. If they are interested (and your first job is to engage their interest) they will make mental summaries of what you are saying, and relate those summaries to what they already know. Then they will try and build a memorable structure, filtering out what they don't need or cannot understand or accept. No one can remember an entire presentation after a single hearing. See if you can remember just one page of your own script after reading it once. For them to understand, accept and remember what you say, you must help them. *Follow a simple structure* (that you share with them) and relate your new ideas to established thinking on your subject.

I'll return to what you could and should do about holding their attention and improving their recall, but first let me show you what happens in every audience. In fact, in every person present. Their concentration span is short, and they drift in and out of what is being said, as this diagram illustrates:

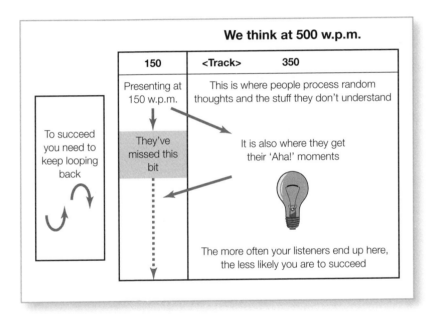

Track 150

We think at an average speed of 500 words per minute or more. *You are on Track 150*, presenting at an average of (say) 150 words per minute, taking the audience from where they are to where you want them to be (dotted arrow).

Track 350

Your listeners' brains have a surplus of at least double that (350 w.p.m.). That surplus capacity is not going to lie dormant. Thoughts will intrude, even if your listeners are trying to give you their full attention. Each time that happens, they will be on Track 350.

If you say anything they do not understand or do not agree with, they will deal with it on Track 350. That is also where they get their 'Aha!' moments, when they get the point you have been making. Each time, they drift onto Track 350.

Of course, they return to join you on Track 150, because you are making the most noise in the room, but each time they have missed a bit of what you were saying.

Typical presentation timeline

INTRODUCTION (Half attention) → **HOOK** (Track 150) → MAP/AGENDA (Track 350, visualising sequence) →

THEME 1 (Track 150) → (intermittently Track 350) → TRANSITION/LOOP BACK (Track 150) → RHETORICAL QUESTION (Track 150) →

THEME 2 (Track 150) → (intermittently Track 350) → TRANSITION/LOOP BACK (Track 150) → RHETORICAL QUESTION (Track 150) →

THEME 3 (Track 150) → (intermittently Track 350) → TRANSITION/LOOP BACK (Track 150)

SUMMARY (Track 350, putting elements in pigeon-holes) → CALL TO **ACTION** (Track 150)

This is, of course, only an example of how attention may drift in and out, and the pattern will vary according to the individual concerned and the occasion. The one certainty, however, is that it will occur, and you will need to plan to recapture attention at frequent intervals.

It happens in daily conversations

Think about the way you listen to someone in conversation. While they are telling you about some incident or experience, don't you often:

- think of something they reminded you of;
- prepare what you want to say at the first available opportunity;
- work out the right point at which to come in?

It's the same thing. You may think you are multi-tasking, but that's a myth. What you are actually doing is drifting in and out of Track 150.

We all do it. It's going to happen when you present, so expect it and plan to cope with it. *The best way is to loop back.* Using your transitions, and referring to the 'Map' you gave at the start of your presentation, remind people of what you have just covered, and tell them what's coming next. That helps them to catch up and make sense of the gaps in what they have heard.

Your transitions should be something like: *'That's the problem as I see it. Let's now consider how it came about. We need to know the causes of the problems we identify before we can find the right solution. So what should we do next?'*

Such a transition simply marks where you have arrived and where you are going next. It adds no new information but it does allow listeners to catch up and also to file away what they have heard from you about the previous section (in this case, it would be defining the problem).

Essential tip

■ Use 'transitions' to loop back and help the audience to catch up.

Power of association

What else could you do to help your audience understand and retain what you tell them? Illustrations help. They help with attention, with understanding and with recall.

The Open University recommends visual techniques for revision: 'Developing visual material can help your recall and also be a quick way to show lots of information. Visualisation helps you remember – two examples are trying to picture where you've left your car in a car park, and thinking about what's inside your cupboards when writing a shopping list.'

Tony Buzan, in teaching *Mind Mapping*, strongly recommends using colours and pictures. He says: 'Colour is one of the most powerful tools for enhancing memory and creativity' [and] 'adding images . . . increases the possible triggers for further associations and recall.'

Essential tip

■ Use visuals and colour to aid recall.

Conditioning plays a big part in the way people listen, just as associations help recall. What is in the mind and memory will act as filters to influence the way we receive new ideas, information or experiences.

Dim the lights and put on a video or film clip and people will go into movie-watching mode and expect entertainment. They may even want popcorn! Some will fall asleep.

If the clip is well made, they are likely to engage their emotions, because that's what happens when they watch a movie. You will then have to be sensitive about the way you speak when the clip is over. It's a bit like the way you might handle someone who has just come out of hypnosis.

If the clip is poorly made, however, you'll find your audience judging it against the professionalism of TV and the cinema. They will not be listening to the message, because their minds will be occupied with a critical analysis of the clip's shortcomings.

Overcoming the rejection reflex

We live in an *Experience Economy*. We judge everything according to the experience it delivers. If we are networking, we decide whether we like the people we meet. When we attend business presentations we judge them according to how good they made us feel. There isn't time to make detailed, reasoned assessments, because we all have too many demands on our attention.

Every day, everyone in the developed nations has huge volumes of data, information and sensory appeals to deal with. An average edition of a Sunday broadsheet newspaper has much more information than all the books available in the 15th century. In 1472, the world's best university library, at Queen's College, Cambridge, boasted just 199 books. You probably have many more yourself.

Today, we are assailed by emails, newspapers, magazines, press advertisements, radio, TV, posters, notices on the railways, advertisements on Tube stations, along the walls of escalators, text messages on our phones, phone calls, Tannoy announcements, traffic notices and more besides. The average supermarket displays something like 40,000 items, and we have to look at them as we pass, to decide if they are what we want.

All this amounts to information shrieking for our attention. The Nobel Prize winning economist, Herbert Simon, explained the effect of this, saying, 'What information consumes is . . . the attention of its recipients. Hence a wealth of information creates a poverty of attention.' We have to develop a coping strategy for all these demands on our attention, and that strategy is a simple one: we switch off. We have been trained to do so.

That is the obstacle you have to overcome when you present. The rejection reflex.

Your audience brings this reduced capacity for attention to your presentation. Think of attention as bandwidth: the ratio between what they can remember and what they will forget. Their bandwidth is depleted.

The way to handle it is to:

■ offer very few items of information at a time;

■ be different, be unexpected;

■ tap into the listeners' basic needs;

■ make them feel good about themselves;

■ use visuals and colour.

Essential tip

■ Think of audience attention as bandwidth, and don't overload.

How audiences process information

The danger in spoken communication is this: what the speaker says may not be what the listener hears. They may both process information differently. Both will be clear in their own minds about what was said and what was meant, but often the messages will not match.

Expect there to be broadly three types of listeners in the audience:

1 Those who are hostile.

2 Those who are passive.

3 Those who want value.

Hostile

The normal distribution curve suggests that there will always be a small percentage of people who will be hostile, and a similar percentage who will love you, come what may. You just need to be aware of potential hostility and prepare for it. When preparing a presentation for a large group I always imagine that only three people will turn up, and sit defiantly in the front row with arms folded. If I can win them over, I'll win them all over.

Passive

In any large audience there will be a number of people who are content to receive what you say in much the same way as they might watch a television programme. They do not have an agenda and either may have been instructed to attend or will be part of the implementation of any decisions that may be taken after your presentation. It would be a good idea to connect with them and capture their interest.

Value

The decision makers will want value. They will be listening for something new, something they can use. They may not adopt your proposition totally, but use it as a jumping off point. It's not necessary to give them the full solution, but rather to inspire insights, and allow them to suggest applications of your proposition that might lie outside your own thinking.

In the early 20th century, Jim Young was the Creative Director of the advertising agency, J. Walter Thompson. In his book, *How to Become an Advertising Man*, he defined five advertising functions:

1 To familiarise.
2 To remind.
3 To spread news.
4 To overcome inertia.
5 *To add a value not in the product.*

There was a TV commercial for the Peugeot 206, in which a young Indian man beat his plain, ordinary car (with the aid of an elephant) into a shape resembling the Peugeot, to give himself street cred. The advertisement won an award. It was not directly about the car, but rather about the perception of the brand, of the style, and of owning a car of that shape and style.

That street cred was a value that was not intrinsic to the product itself.

How large groups and nations process information

There are some overriding influences on the way we process information. Being common to all groups, they are what I'd call 'collective' behaviour, and they may characterise a whole nation's way of thinking, and carry over into your audience.

Collective behaviour is when groups of people respond in the same way as one another, even spontaneously demonstrating some shared impulse. Think about football crowds that suddenly start singing the same song *in the same key*!

> *A presentation audience can similarly turn for or against you, either 'getting it' or deciding they do not understand or agree. It is therefore a good idea to pick the 'friendly faces' in your audience and quickly get them nodding in agreement. It will help the others to follow.*

It's worth considering a nation's way of reasoning, as it may influence the group behaviour of your audience. The reasoning styles fall into two broad camps:

1 **Particularist.** These people look at each situation on its own merits. They include Egyptians, French, Greeks, Chinese, Portuguese, Mexicans, Singaporeans.

2 **Universalist.** These people look for universal rules or laws to govern their decisions and behaviour. They include Australians, Brazilians, Canadians, Finns, Germans, Britons, Swedes, Americans.

Dealing with facts

Many presentations fail because they are overloaded with facts and figures. Many people close their minds to figures, and few people can remember many new facts. There is a popular brain game called Sudoku. It consists of placing numbers 1 to 9 in a grid, so that all nine numbers appear in every vertical and horizontal line, and also in every one of the nine boxes. That description is actually more complicated than the game itself, which does not require any skill with numbers. It could be played with nine symbols of any kind. Yet I have met people who refuse to consider playing the game simply because they are phobic about numbers.

Just imagine how they would respond to numbers in a presentation.

Essential guide to handling facts

To repeat something I wrote earlier:

■ facts and figures are neutral;

■ facts must be interpreted to become information;

▶

- information must be understood to become knowledge;

- knowledge must be filtered through your point of view to become your wisdom.

FACTS (interpreted) → INFORMATION (understood) → KNOWLEDGE (filtered) → WISDOM

Your listeners want your wisdom, not your facts.

Why you must be succinct

Television is the most common medium of communication these days. It has changed the way in which information and entertainment are delivered, and trained the public at large to expect information in a certain way. What people want is:

- instant gratification;
- satisfaction without effort;
- pre-digested news;
- to know what they *need* to know to keep up.

Even computers mimic television, with video clips about everything. YouTube delivers presentations, demonstrations, entertainment, and more. Some of it is rubbish, a lot is not. It sets the standard against which your presentation will be judged. Moreover, *the internet and YouTube have made your audience better informed than ever before,* and their expectations will be based on their online searches and experiences.

Think of a news story on TV. Often the story starts with the news anchor in the studio, cuts to an 'expert' who then refers you to a reporter at the scene, who may interview someone affected by the story. Each person in the chain makes a brief contribution, but the scene changes each time, adding to the visual input. Then it is on to the next story. World news in 20 minutes, and never a moment's silence, never any 'dead air'.

Meanwhile, there could be domestic sounds in the viewer's background, perhaps from the kitchen or from family members. It takes only 15 per cent of the brain's capacity to understand and process language, leaving 85 per cent to deal with interruptions. That trains people to listen with half an ear. That inbuilt inattention is the obstacle you face when you present, as well as the expectation of information TV-style.

Essential tip

■ Be succinct. Follow the example of TV news bulletins.

Body language

In a business context, information is generally not received passively. There is a need to evaluate it, asking, 'What does this mean?' and 'Can I use it?' Even more importantly, 'Do I trust the person giving me this information?' This is when body language becomes important.

Your listeners read body language at least as well as they understand the words you speak. *Words are cerebral, body language is visceral.* It is not enough to say the right things. You have to demonstrate your commitment to your message through your body language.

Unspoken language

The key question to ask yourself (while you watch a video of your rehearsal) is this: 'Are you delivering the words or the message?'

This matters even more when you are presenting one-to-one. You can't fake sincerity, and when you are presenting to just one person, your style, your voice and your body language will be quite different from when you present to a large audience.

Try this. Choose a topic that you know well, and speak about it on camera for three minutes. Then write a one-page script on the same topic and learn it by heart. As soon as you know it, recite it on camera, without reading from the script. Now compare the two versions. Can you see and hear the difference? You may be more fluent in the second version, because you have prepared the words, but do you sound more convincing?

If you remember only one thing from this book, let it be this: *you must present in a way that connects with the way people listen.* It is the most effective way to get and retain attention and to make sense of WII FM.

Essential tip

■ Speak heart to heart, not brain to brain.

We'll develop this further in the next chapter, in which we consider *delivering your presentation*.

Summary

- Tips for overcoming nervousness
- How people listen
- Typical timeline for presentations
- Why you need to deliver a good experience
- How to deliver the facts

Delivering your message

In this chapter

- Power in delivery
- Appearance and charisma
- The way you sound
- Nine essential microphone techniques
- Pitch, pace and pauses
- Connecting with the audience
- Gestures and movement
- Presentation essentials

You have thought about what you want to say, and why, and you have drafted a well-structured presentation, complete with visual aids. What is there still to do? You have to do justice to all that has gone before. Your delivery has to achieve your objective.

In this chapter we shall consider the many aspects of platform mastery, but mostly we shall be looking at salesmanship, because you must 'sell' your message. It's up to you.

Effective delivery depends on:

- your voice;
- your commitment;
- knowing what you want the audience to accept;
- taking charge of the platform;
- connecting with the audience.

Before I deal with each of these points, I'd like to make two important observations about public speaking or presenting. First, you don't have to be a 'born speaker' to do it well, and second, a speech or presentation is much more than amplified conversation. To do it well, you must have something to say, something that you really want others to hear, something that belongs to you, which comes from your heart.

If you accept that, you cannot believe that it is all right to deliver a presentation in the way that most people do – reading a script word for word, as though it were enough to deliver the words alone. You must have a point of view to put across, and that is best conveyed by the conviction in the way you speak. Reading a script will tend to diminish that conviction.

Try this: rehearse your presentation in front of a camcorder and play it back to a couple of people whose judgement you trust, and who are not afraid to tell you the truth.

After they have watched your presentation, ask them these questions:

- What was the core message?
- What was the sequence of the argument?
- Did they accept/agree with what was said?
- What would they do differently as a result?
- What did you do well?
- What should you do differently?

Power in delivery

There is potentially great power in the way you put your point across. Many presenters fail to use it because they don't understand the effect they can have on their listeners. Let's consider just two examples.

Example 1

Let's think of a person making a presentation about water coolers (although it could be almost any product). In a typical presentation, they would be climbing up the side of a mountain, hoping to reach the summit, where they get the order. It's a battle against the resistance, real or imaginary, of the prospect. The prospect, in turn, is also viewing the scenario as a confrontation, a battle for a portion of their budget. And why? Because the presenter has created that scenario by saying, 'I can offer you these benefits' and the buyer is adding up the benefits they hear, to weigh them against the money they would have to spend to get them. In their mind, they are examining the purchase, to see if it is a good buy.

Now imagine a small shift in thinking. The presenter no longer speaks of the benefits they can offer the prospect, but speaks instead of the *benefits the prospect can offer to their staff*, and to their future staff, as one of the benefits they can build into their recruitment package. Now the sales person is no longer an adversary, but rather a *potential colleague*, and the discussion is no longer about a purchase but rather about enhancing the working environment.

Example 2

When Amdahl, a multinational computer manufacturer, had a marketing re-fit in Europe, their advertising agency gave them a new slogan: *The infrastructure company.* But when the company's nine vice presidents from around Europe assembled in England to discuss the marketing strategy, they discovered that the slogan did not really express the way they did business.

When they asked my opinion, I pointed out that they were focusing on the technology they provided to their clients. I said, 'Why not look, instead, at the way your clients deal with their customers, and focus on helping *that* process?' I offered them the alternative slogan: *Amdahl makes it happen.* They agreed that it was a better slogan to describe the benefit to their clients.

Essential tip

■ Base your presentation on helping the audience do whatever they do.

The mind instructs the body

Another important consideration is the effect of what you say on you, if you believe in it yourself.

In the 1950s there was a scare about 'subliminal advertising' – messages flashed on cinema screens so fast that you were not aware of them, but which affected your motivation to buy. At the time it was considered an unfounded rumour, but banned all the same. Recent research has confirmed that it could be true after all.

The messages sent to your brain have a direct affect on your body and even your state of being. Try this little exercise now. Stand up and extend your arms to the side, then slowly twist to one side, watching your extended hand as you go, and making note of where you are pointing when you go as far as you can without straining. Return to the front and shut your eyes. Now imagine yourself repeating what you have done. When you reach your previous limit (in your mind's eye), go a little further, and a little more, then return to the front. Now physically repeat what you did originally, and see what happens when you reach your original limit. You'll go right past it.

The mind instructs the body. And what you say to your audience will have the power to change them. In fact, quite often they will want to be changed. They will be hoping for some solution to an existing problem, some new insight, something they can add to their own portfolio of skills.

How strongly does the mindset affect behaviour? Well, imagine how you'd feel if your business ran into a severe cash flow crisis. Perhaps you'd be reluctant to take a risk of any kind, and you'd draw in your horns. Then suddenly there is a large injection of cash. What would that do to your state of mind and your approach to risk? A change in your circumstances alters the way you think and act.

If you create in your audience the right level of expectation or anxiety about a problem or need, they might hope for the solution from you. Don't disappoint them.

Essential tip

■ Create hope in their minds and they will respond to you positively.

Appearance and charisma

Let's return to the beginning. How do you look on the platform? (For the sake of simplicity I refer to where you stand as the platform, whether or not there is actually a stage or platform.) I once saw a speaker whose script and technique were both very good, but he looked like he had slept in his suit. A couple of people in his audience remarked that he was good enough to make a demonstration speech, but they would not wish to follow him or accept his guidance because of the way he dressed.

Even now, a number of years later, I always visualise him in that rumpled light grey suit. He will never get another chance to make a first impression on me. Does it matter how you dress for a business presentation? Only if you want to be taken seriously.

Because you will judged within the first ten seconds of rising to speak, it will help your cause if you take the time to think about how you look. In my opinion, you should aim to match the best dressed person in the room. Be flamboyant, by all means, but ask yourself if you really want people to be concentrating on your appearance rather than on what you are saying.

Charisma is something else. If you have it, great. If not, work on developing it.

Essential guide to charisma

To develop charisma, you need to be:

- positive and upbeat;
- confident and cheerful;
- lively and energetic;
- knowledgeable, and not just about your own topic;
- polite and patient;
- able to relate to the audience;
- committed to a clear point of view.

Be committed to helping the audience

In Chapter 4 I wrote that you must have a message that you feel passionately about. That's essential if you want your presentation to succeed. But there is a qualification: your passion and commitment should be about the benefit to your listeners. If you simply want to get something off your chest, that could come across as self-interest, and your listeners will switch off.

Your audience may include people who either don't know you at all or who know you only slightly. You are standing before them, telling them something they did not know before, expecting them to understand, accept and act, all on a single hearing of your message. That's a tall order, isn't it? It requires trust.

Trust begins when your audience can see that you are genuine, and committed to helping them. Trust is also easier to gain when the commitment you want from them is small and easy to give, so consider a multi-layered solution. As each layer works, their trust will be increased.

The way you sound

Have you ever heard your recorded voice? Did you like the way you sounded?

The first time people hear themselves on tape, for example, they wince. Their voice is not as attractive as the sound they are used to, and it may even lack energy, warmth or friendliness.

The reason why it sounds less attractive is that it lacks the usual resonance. When we speak, we hear the resonance from within our own skulls, a resonance that is not transmitted to the outside world, so the recorded voice sounds flatter to us when we hear what others hear. If we speak in a matter-of-fact conversational tone, we can sound dull and lacking in energy or warmth.

Fortunately there is a single solution to both energy and resonance. It isn't hard to make the change, but it does require a little effort and practice. As in all things related to effective presentations, practice is key.

Essential tip

■ Listen to the way you sound and change it if you sound dull.

You *can* change your voice

Changing your voice involves two things. One is physical, the other mental. The former takes a little longer and involves a certain exercise which I'll come to in just a moment. Let me focus first on the way your mindset affects the way you sound.

If you are preoccupied, self-conscious or in a businesslike frame of mind, you could sound unfriendly on the telephone. Think of the last time you spoke to your spouse or partner while they were on the train. How did they sound? Was it as warm and friendly as usual? Or did it sound restrained, cautious or even depressed?

Consider how you respond to a phone call from a stranger. If it sounds like someone wanting to buy from you, how do you respond? And what happens the moment you realise the caller wants you to buy from them?

Now imagine you have made a discovery that overturns conventional thinking about something or other. When a journalist calls to interview you, how will you sound? Excited, authoritative, confident in your expertise? Of course. So speak with confidence in your ability to offer value to the other person. It will add energy and excitement to your voice.

One more thing about the way you sound. On the day I was writing this, I attended a meeting in London. One of the people there spoke about his business, but I had difficulty receiving everything he said because his voice lacked resonance. But he also spoke in jargon and he had a tortuous way of framing his sentences.

No one else in the meeting understood what he had said. The chairman even told him, jokingly, that it sounded like Klingon. The winning presentation formula must be simple words, simple sentence construction and a clear, resonant voice.

Essential exercise for vocal resonance

Here is a simple exercise to improve your vocal projection and resonance. Hum and feel the vibration in your face. Then just drop your jaw to open your mouth and you will be making an 'Aah' sound. Open and close your mouth so that you are going 'Mmm . . . aah . . . mmm . . . aah . . .' but always try to maintain the vibration in your face. This little exercise will bring your voice out of your throat and bounce it against your front teeth, which will improve your resonance. A resonant voice is more attractive to listeners and carries further, making it easier for people at the back of the room to hear you.

Using a microphone

Even if your voice has resonance, there will be times when you should consider using a microphone. If the room is big enough and you have a large audience, if a microphone is available, take it. You will be able to employ more vocal variety, avoid swallowing the ends of your sentences, and you will dominate the room with your voice.

I believe you should use any aids that are available to you, but that you need to master the appropriate techniques to use them effectively. Time without number I have seen people come on stage from the audience (for example, at awards ceremonies) and lean into the stand microphone, even though they may have been told not to do so. Not only is it unnecessary, it also detracts from your authority and personal presence if you do so.

Nine essential microphone techniques

1 Don't lean into the microphone. Raise it so that you can stand tall and speak as though the mic were not there.

2 Position the mic about nine inches away from your mouth. With an open palm, place the tip of your thumb against your lips, as though playing a trumpet. Your little finger should touch the mic.

3 Find the sweet spot. There is a position where the mic picks up and delivers your voice best. You can hear it yourself.

4 Don't shout. Let the mic do the work for you.

5 Always position the mic below your mouth. If you place it higher, it will block off part of your face, and when you look down at your notes your voice will disappear.

6 Don't get too close, or the mic will pop with every 'p' and 'b' plosion, and hiss on every sibilant 'ssss' and 'sssshhhhh'.

7 If you have a strong voice, when the sound engineer asks you to try for a sound level before the meeting starts, deliberately speak with less power than usual, otherwise the sound engineer will set a volume level that's too low, and you could lose impact.

8 Always keep the mic in front of your chin. If you turn to the left or right while speaking, imagine that a chopstick links your chin to the mic, forcing you to move your whole body so that the mic is always directly between you and the people you are addressing.

9 If you wear a lapel microphone, get someone to tell you, in rehearsal, if you tend to speak more towards your right or left, and wear the mic on that lapel. Check that it is firmly clipped in place and that your movement and gestures do not cause any part of your clothing to knock against the mike, as that can irritate the audience.

Take charge of the platform

Once again, let me remind you that you are being judged the moment you are seen, even if someone else is speaking before you, and people are making up their minds about you. Decide how you want to be perceived, as a dominant speaker or a humorous one, and prepare your opening to establish yourself in that mode.

Moving around is a good idea, so long as you move with a purpose. Just striding around can be distracting.

Three good reasons to move

1 To help the interpretation of what you are saying.

2 To enable everyone in the audience to see you better.

3 To check that all heads are following you.

1 Enhance interpretation. Imagine someone delivering a presentation while standing stock still, arms by their sides. Wouldn't look right, would it? You'd expect some animation, gestures to illustrate and emphasise certain points, even a few strides to either side. After all, the presenter is the presentation, and it's a live performance, not a still photograph. Apply the same thinking to your own performance.

2 Improve your visibility. It could be that not everyone can see you clearly, so make it easy on them by moving and standing in different places from time to time.

3 Check they are following. Watch the audience's heads as you move. If they follow you, they are paying attention. If not, you will need to re-connect with them.

Pitch, pace and pauses

These are the variables that add variety and interest to a speech or talk.

Pitch

The key in which you speak, or the main note that is heard. Find the right starting note, so that you sound authoritative and natural, and feel comfortable.

Beware the tendency to go too high. The larger the audience, the greater the temptation to strain and raise the pitch.

Practise speaking into a tape recorder, playing back with the volume turned down, to decide if your voice sounds attractive or monotonous.

You are listening for the tone, not the words. Try recording the same passage at several different pitches. Try switching to a different pitch (usually lower) mid-speech, for dramatic effect.

Pace

The speed at which you speak should also be varied. The ideal speed is between 140 and 160 words per minute. The more energy you put into your speaking the slower you will be, even if it doesn't feel that way.

Speaking too slowly and deliberately sounds pompous. Speaking too quickly can cause words to be swallowed, and for them to run into each other. To give the impression of high pace, without losing clarity, hit the end consonants of your words.

Pauses

Take your time and do not gabble.

Use the pause for dramatic effect, and to allow your point to sink in.

You should pause:

■ at the start, to get attention;
■ before a significant piece of information;
■ before the punch-line of a joke.

Variety keeps your listeners interested and signals your willingness to communicate.

Involve your audience

A business presentation is not a lecture. Your audience should not be expected to sit silently while you ramble on to the end. Get them involved, make them move, encourage questions.

In fact, ask them questions. Always ask rhetorical questions such as, 'So what's the answer?' Such questions do not require an answer, but they are the kinds of questions that may be in the minds of your listeners, and when you ask and then answer those questions, you make your listeners feel they have participated in a dialogue.

For variety, ask questions that do require an answer, such as, 'Who here has had such an experience?' Always have a follow-up question, depending on the response you get. If no one raises a hand, be prepared to ask, 'How would you react if you *did* have such an experience?' And if hands do go up after the first question, ask them to describe what happened and how they felt about it.

If the audience have been sitting for any length of time, ask them to stand and stretch, or give them an activity that requires them to move around. Their brains will clear and they will listen more attentively.

Essential tip

■ If possible, get your audience to move. It will keep them alert.

Connecting with the audience

Delivering a presentation should not be a one-way street. There are at least four distinct levels of communication taking place:

1 Audience's needs.

2 Your purpose.

3 Content.

4 Outcome.

Now let me explain each in turn:

1 **Audience's needs.** Remind yourself of why they are there, and what they might want to gain from listening to you. What can you add to your presentation that is specific to the listeners? What's the first thing you are going to talk about? Remember, it should not – must not – be, 'This is who we are and this is what we do'.

2 **Your purpose.** Remember that you should be aiming to make some change in the thinking, attitude or behaviour of your listeners. What is that going to be?

3 **Content.** A presentation should not be all froth. It must have substance, something of value that is credible and verifiable.

4 **Outcome.** What do you want the audience to do when you have finished talking? Be clear about it and remember to tell them what to do.

Now consider who your audience are and how they were taught to learn. In some cultures, it is the custom to take notes, so you will have to make it easy for them to do so. Other nations, such as the Japanese and the Finns, are liable to listen in total silence, ask no questions and give you no feedback. The Nigerians and Koreans are very participative and will answer even your rhetorical questions.

Even if you never make presentations to people from other nations, it is worth bearing in mind that people may listen and learn very differently from you.

Now consider what happens in the first few seconds. It's what some people call the 'sniffing period'. Think about two dogs meeting for the first time. They sniff each other and decide 'friend or foe'.

A smile and a strong voice are a good combination to start with. If you are confident about a humorous opening, especially if you make it relevant to the occasion, use that (but not with Germans, Finns or Japanese). Just DO NOT TELL A JOKE. Few people can tell jokes well, and most jokes are so contrived as to lose you the respect of many listeners.

Gestures and movement

Gestures help to illustrate your message, indicating emphasis, punctuation, size and position. They are also about rhythm, energy and passion, but not about beating time!

It is essential to develop good-looking gestures and rehearse them in front of a mirror or video camera until they look elegant, smooth and perfectly natural. You should also move.

Ten essential tips about gestures and movement

1 Do not turn your back on your audience.

2 Do not pass your hands across your face.

3 Make generous, open arm gestures.

4 If you are walking about, stop and anchor your feet when you want to make a telling point.

5 Move with a purpose and plant your feet, or you will lack authority.

6 Do not cross your legs, either when standing or when moving from side to side.

7 Avoid pointing your forefinger – it's too aggressive!

8 Remember that certain gestures have different meanings in other countries. A 'good' gesture in one country may mean something rude in another.

9 Stand tall. Touch the ceiling with the top of your head.

▶

10 Smile. It's hard to resist someone who seems pleased to see you. Start the process by thinking of some happy incident that always makes you want to smile. Your face muscles will start to lift, and your spirits will as well. Pick a friendly face and exchange smiles.

Presentation essentials

1 Message. This is the essence of your content, the core value you are offering, the interpretation that you put on the information you impart. It is your wisdom. As I have said before (and will go on saying) if you want to transmit information, send an email. There has to be something more.

2 Messenger. Why is it you delivering the message? Is it your own thinking on the subject, or can the audience equally well hear it from anyone else? For it to make an impact, it must come from within you, and belong to you. The message must be linked to you and identified with you, and you must want your listeners to make some change as a result.

3 Method. However good your content, however powerful your message, you need the right skills to put it across effectively, to stir your listeners' emotions and make them want to accept and act on what you say. Most people need coaching or training to develop those skills.

4 Rehearse, rehearse, rehearse. You must know your stuff. Some of the best presenters in the world practise: they seem relaxed and informal, but they have rehearsed every vocal inflection, move and gesture.

5 Have a printout of your slides, clearly numbered, to give you full control. It will enable you to jump to any slide, simply by typing in its number and pressing Enter.

6 With a large audience, use a microphone. Give yourself every opportunity to be heard clearly by everyone. Those who miss a few words will soon switch off.

7 Check that everyone can see your face. Where is the projector located? Your face is a major visual aid. It tells the audience how you feel about your content.

8 Position your flip chart where you can write on it without turning your back, and practise walking to and from the flip chart while facing the audience.

9 Do not look at the slides on the screen. As soon as you do that you are telling the audience that you are merely the acolyte and not the presentation itself.

10 Refer to your 'Map' and use transitions between sections. Remember, you are taking your listeners on a journey, and they need to know where they are, from time to time.

11 Maintain good eye contact, holding it with specific individuals for at least three seconds, so that the audience will feel involved and know that you are speaking to them and not at them.

12 Give out handouts at the end, not in advance, otherwise you could find the audience reading your handout and not listening to you.

13 Speak with energy, but not too fast. If your natural speaking speed is fast, hit your consonants even harder than usual. You will still feel as though you are going fast, but will actually slow down considerably.

14 Follow the sequence of persuasion, getting agreement at each stage. Remember AIDA: Attention, Interest, Desire, Action.

15 Make sure you keep driving home your core message. It's what you want people to carry away with them and remember.

These reminders will help you to deliver an effective presentation, so long as you also observe the rules for preparation.

In the next chapter we'll discuss some *advanced concepts*, such as the emotional journey that you take your listeners on, and how to plan and manage that to good effect.

Summary

■ The sales pitch that's like climbing a mountain
■ The mind determines what the body does
■ Guide to developing charisma
■ Changing the way you sound
■ Taking charge of the platform
■ The right gestures and movement
■ Presentation essentials

part

3

Reviewing and coaching

10

For experienced presenters

In this chapter

- The presentation wheel

- Six different kinds of presentation

- The right language

- Oratorical devices

- The emotional journey

This book is about persuasive business presentations. It is about the process of changing the way others may think about the issues you want to discuss, so that they are prepared to take the action you propose. It involves providing evidence and a well-structured case that builds on what the audience already knows and takes it on a notch, because change is not easily accomplished in giant strides. Part of the process is meeting the expectations of your listeners.

This chapter is about a level of understanding that goes beyond the mechanics of putting together a presentation and delivering it competently. It is about managing the emotional journey created by your presentation, and about the language and techniques that skilled orators use to reach the hearts of their hearers.

What follows is a wheel diagram with ten spokes. They represent the *ten most important elements* of a business presentation. The diagram provides you with a means of measuring how closely your own expectations might match those you might be addressing.

Make a photocopy of the diagram and mark each element out of ten for importance, in your own opinion, by marking a dash on the spoke. Zero is at the centre of the circle, and ten is on the rim. When you have marked your ten scores, join up the dashes, and you will see a shape that represents your preferences.

Next, make another copy and give it to someone who is similar to the kind of person who would be in your audience. Ask that person to mark their own preferences, just as you did. Now compare your diagram with theirs. That will give you some idea of how close or far apart your expectations might be.

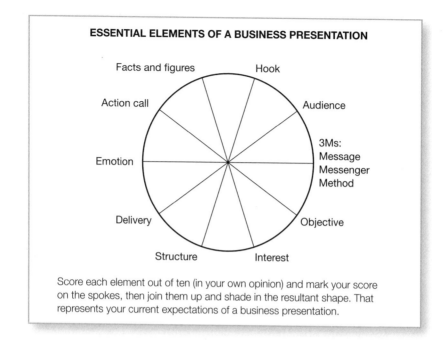

ESSENTIAL ELEMENTS OF A BUSINESS PRESENTATION

Facts and figures — Hook — Audience — 3Ms: Message Messenger Method — Objective — Interest — Structure — Delivery — Emotion — Action call

Score each element out of ten (in your own opinion) and mark your score on the spokes, then join them up and shade in the resultant shape. That represents your current expectations of a business presentation.

The ten elements

To help you decide on your scores for the ten elements on the wheel, here are some brief explanations of those elements. A simple mnemonic to help you remember them is **HAMO IS DEAF**.

H is for the Hook

As explained earlier, this does the same job as the headline in a press ad. It is something that you do or say to grab the attention of your audience. After all, it is obvious that you cannot make your pitch without the engaged attention of your listeners. Thanking people for being there is not the best way to open. You can thank them later.

Think about the way films and even TV dramas open these days. They go straight into the action, setting up an incident or an encounter that will determine the way the plot develops, and bring up the titles and cast later. In other words, they open with a hook. Remember, the makers of films and TV dramas know a thing or two about connecting with their audiences. You can use a prop or you can say something unexpected, just as long as you do something more impressive than a seamless move into a PowerPoint slideshow.

The hook is the A in AIDA – it gets Attention.

A is for Audience

Your presentation *must* be relevant to the audience, to their needs and interests. This means making the effort to find out their concerns, and their business objectives. What anxieties do they have that you can enlarge and relate your message to? Do they have any cultural or social influences that can affect the way they respond to you? Do you have to modify the language you use to connect well with them? What can you find out about their pressure points, their tolerance levels, their attitude to you and your company?

If you do no research and encounter hostility or resistance halfway through your presentation, you may find it hard to recover. It's better to find out in advance and prepare for it. In the TV series *Dragons' Den*, many would-be entrepreneurs are blown away because they do not prepare adequately, and do not anticipate the 'So what?' questions that the Dragons like to ask.

M is for M x 3

The 3Ms are **Message, Messenger, Method**, and are more fully explained in Chapter 9. If you are going to make a presentation, you must have something to say, something you want your audience to hear and act upon. That's your *Message.*

The second M is about you, the *Messenger.* Why is it you giving this presentation, or why do the audience need to hear the message from you? It needs to belong to you, because it expresses your own take on the ideas or information.

The third M stands for *Method.* This is about technique. You do need to develop the communication skills that enable you to make the most of your message.

O is for Objective

Why are you making the presentation? More importantly, what do you want to achieve, and what do you want to happen at the end of the presentation? It is highly unlikely that you will prepare and deliver a business presentation without some objective in mind, but will it be the right one?

Many a person will say, 'My objective is to promote my business.' Is that good enough to tick the box? I have often heard the American motivational speaker, Les Brown, say that the problem is not that people aim too high and miss, but that they aim too low and *hit.* So it's really worth taking a close look at what you can achieve with every presentation you make, and make sure you aim high enough.

I is for Interest

The AIDA graph in Chapter 6 clearly indicates the importance of capturing interest and building it to the point of desire. You do that by following a structure, by clarifying the relevance of all your facts, by using stories to make your point, by ensuring that your audience remains engaged. This is about the body of your presentation, and should follow a three-part structure, such as Past/Present/Future or Problem/Cause/Solution. No matter how many things you want to talk about, group them into three streams of argument or reasoning.

S is for Structure

There are two quite separate parts to structure. The first relates to the sequence of the content, as suggested in the previous paragraph. It's the physical organisation of the content. If it is not structured, if it does not follow a discernible pattern, it could be hard to follow and even harder to recall. A planned structure keeps the presenter on track and helps the audience follow what is being said.

The other part of structure is about the emotional journey. If your objective is to take your audience from where they are to where you want them to be, it will be a process. It is highly unlikely that you could just announce your proposition and get it immediately accepted. You will have to warm up your audience, engage their emotions as well as their minds, and take them gradually up the emotional scale until they say, 'We'd like to have that.'

That's the point of Desire – the D in AIDA.

D is for Delivery

Think again about the point made under 3Ms (above) about why it is important to receive the message from you. The presenter makes a difference to the success of the presentation. A vital element in the delivery is conviction. If you believe in your proposition and in its value to your listeners, you should speak about it with conviction and commitment, and that can make all the difference.

Several organisations were involved in a joint venture that came under attack, and they were required to justify their actions. One man organised the defence, with one director from each of the member companies. He himself was going to act as chairman, calling each up in turn.

▶

While rehearsing them, I realised that he was the best presenter of them all, although he had an easygoing, avuncular style. That was because he believed in the rightness of their cause, and in the original project. I switched him to lead presenter and the case was won. Delivery matters.

Emotion and passion

Here again, the emphasis on emotional appeals will vary according to the audience and topic (and also the country). What importance do you give to emotion in the content and passion in the delivery?

A is for Action call

As I said before, you need to drive your presentation towards some action to implement the change that you are aiming to bring about. If you leave it to your listeners to make the decision themselves, you could have a very long wait! If the presentation gains agreement at each step along the way, it is logical to ask for commitment and to tell your listeners what to do next. Don't hope to play it by ear, plan it.

This is the final A in AIDA.

Facts and figures

There will usually be factual content, supported by evidence. This element will be more important in some presentations than in others, and that will also apply to certain audiences. In general, how important is the factual content in the presentations that you deliver?

Essential tip

■ Even under different names, the AIDA elements of persuasion work.

There are, of course, other elements that contribute to the success of a presentation, such as:

■ **Word pictures and stories.** Would you illustrate your presentation with stories, and by creating word pictures? Does it help make your message more understandable?

■ **Oratory**. How important is it to use oratorical devices such as repetition, groups of three, and the techniques of Barack Obama and other powerful speakers? (This is developed further a little later in this chapter.)

■ **Relationship building**. Is it necessary, in your view, to consider the presentation a part of the process of developing a relationship, or is it enough just to transmit your message and push for a commitment?

■ **Respect**. Your presentation may be trampling on long-held beliefs, or contradict what someone in the audience may have said, or even run counter to their business practices. Is it necessary to take account of hurt feelings or seniority?

■ **Humour**. Do you believe it is essential to use humour in a business presentation? Are there situations in which you would not use humour? What about telling jokes? Are you the kind of speaker who can get away with it?

Six different kinds of presentation

Although some of the rules apply to all business presentations, we should recognise that there will be differences according to the nature and purpose of the presentation. These are the main kinds:

1 **New business.** This is a sales pitch. It needs to follow the sequence of persuasion, with particular emphasis on addressing the needs of the audience. If treated as a problem-solving presentation, it has a greater chance of success. It should therefore play down the credentials element and trumpet blowing, although testimonials could feature as evidence. The 'buyer' or 'prospect' should be encouraged to speak, as you cannot solve problems unless they have been stated or confirmed by the other person.

2 **Persuasive.** There are other kinds of persuasive presentations, for example when you are looking for a vote in your favour on a matter of policy or when there is more than one point of view on a central issue. Although the context is competitive (your stance against the rival one) you should avoid criticising the alternative point of view, as that will often alienate those who are still undecided.

Essential tip

■ Avoid criticising. It turns people off.

3 **Motivational.** This type of presentation usually depends heavily on the presenter's charisma, and will therefore not require many slides (if any). This could be a leadership statement, a rallying cry or a call to arms, for example if your company is battling its way out of a recession and you want the team to fall in behind you, or if you are a sales manager calling for an extra push for sales. Often an organisation will bring in an outside speaker who specialises in motivation.

4 **Entertaining.** After-dinner speeches and convention keynotes can be either motivational or entertaining, or both. Their purpose is to lift the spirits and add goodwill to the occasion, and they will therefore be quite informal. Toasts fall into this category, in which the guest of honour, for example, might be teased.

5 **Ceremonial.** These are presentations for special occasions such as retirements, funerals, promotions or the award of some honour. They will be about the personal qualities of the person being remembered or honoured, and be briefer than some of the other kinds of presentation.

6 **Instructional.** These presentations are delivered either as lectures or discussions in which the presenter acts as facilitator. The material (e.g. slides) will be shown and explained, usually one step at a time, and a discussion will follow. The presentation could be to deliver the findings of a research project on which a decision has to be taken, so there will be an element of persuasion, although its pace and timing is likely to be extended.

The importance of the right language

See if you can make mental images for the following words:

Happiness

Success

Hunger

Achievements

Put them into sentences, and see if that works any better:

- The pursuit of happiness is a noble goal.
- We are looking forward to a successful year.
- Hunger and disease are common in Africa.
- You should be proud of your achievements.

Those are common enough sentences, often used in presentations and in written documents, but what do they do for you? What images do they create in your mind, and do they press your emotional buttons? Try these for size:

- Getting out of bed each morning, greet the day with a welcoming smile on your face.
- Our aim is to double last year's sales and double the bonus payments too.
- Imagine a child who hasn't eaten for ten days, grazing on grass.
- Everyone in the company and beyond would like to copy what you did.

The difficulty with abstract words like happiness and success is that they usually do not create mental images that the listener can relate to, and that reduces their impact in a spoken presentation. The text that's written to be read is not the same as the text that's written to be said. You can always go back over a written text and think about its meaning. If the spoken text leaves you wondering about its meaning, you'll find yourself on Track 350, and not listening to the presenter.

Here are a couple more examples of text that's hard to understand. The first is from a book about getting ahead in business, while the second is from a book on communication skills(!).

1 The appeal of the strategies and portfolios approach lies in its ability to wed the underlying strength of strategic planning to the real world limitations on management control in marketing today.

2 The implications of identifying underperformance and having individual remedies applied has undertones which could conceivably be interpreted as sinister or threatening to staff.

These are by no means the worst examples of language that you will hear in scripted presentations that are written by people who do not understand the difference between the spoken word and the written text. But they are cumbersome and even difficult to say. In fact you would not speak like that in conversation.

As reported in Chapter 2, Peggy Noonan, speechwriter for US presidents, said, *'You must be able to say the sentences you write.'* Amen to that.

Essential tip

- The text that's written to be said is different from the text that's written to be read.

How skilful orators stir the emotions

A good speaker uses a number of linguistic devices, or figures of speech, to enhance the impact of a speech. The most common are **metaphor** and **hyperbole**. The British use metaphor all the time: the boot was on the other foot . . . it was a mountain to climb . . . health warning on the Budget . . . sailing close to the wind . . . pressing people's buttons . . .

Hyperbole is simply magnification of the truth, exaggeration for the sake of effect. When we describe something as fantastic, immense, incredible and so on, we don't mean it literally. These powerful words are used to signify something on a grand scale, but also our own reaction or response to them. If we call an event fantastic we mean it affected us as much as a supernatural event.

In addition to these two, there are other figures of speech that can lift a speech out of the ordinary. President Obama, one of the best orators of our time, favours the **tricolon**, which is the use of three phrases building to a climax.

Obama uses several oratorical devices, including alliteration (e.g. *long live liberty, levity and leprechauns*), rhyme (*while I encourage levity, I more admire brevity*) and iambic cadence (*I wish I was in Dixie*). For the avoidance of doubt, none of these examples is from any of Obama's speeches. However, here is an example of the tricolon that he did use:

> *'If there is anyone out there who still doubts that America is a place where all things are possible; who still wonders if the dream of our founders is alive in our time; who still questions the power of our democracy, tonight is your answer.'*

The tricolon was also used by Abraham Lincoln:

> *'With malice toward none, with charity for all, with firmness in the right.'*

Obama also used the extended tricolon to great effect. Notice how he combines it here with the ascending structure of 'not that . . . but this . . . and this above all':

> *'A new dawn of American leadership is at hand.*
>
> *[1] To those who would tear this world down – we will defeat you.*
>
> *[2] To those who seek peace and security – we support you.*
>
> *[3] And to all those who have wondered if America's beacon still burns as bright – tonight we proved once more that the true strength of our nation comes not from the might of our arms or the scale of our*

wealth, but from the enduring power of our ideals: democracy, liberty, opportunity, and unyielding hope.'

The first two sentences set the scene for his declaration of principles. It works because it clears the decks by dumping what doesn't fit, then reinforcing what does, before adding the new. It prepares the listener's mind for the message. In a business context, you might say something like this:

1 *To those who think we have fallen behind our competitors, we will surprise you.*

2 *To those who have remained loyal to us, we will reward you.*

3 *And to all who wonder if we are still competitive, we say this – we do not need to chase every new fad or fashion. We are about to announce a development that will amaze you, one that places us at the cutting edge of our industry, but which remains consistent with the values that have lain at the heart of the way we do business.*

There are just three more rhetorical devices I'd like to share with you here, although there are others. These are the three you are most likely to use in speeches or presentations, to generate an emotional response. Don't be put off by their Greek names; they are powerful tools for any presenter:

■ **anaphora** (a-NA-fo-ra);

■ **epistrophe** (e-PIS-tro-fee); and

■ **symploce** (sim-PLO-see). All three relate to the use of tricolons – three statements that build to a climax, as in the examples above.

Anaphora

Repeating a phrase at the beginning of each sentence, e.g. It's a disaster when a government shoots its own people; it's a disaster when a business stops trying to be the best and settles for good enough; but it's a disaster of the greatest magnitude when an individual quits because he or she has been knocked down once too often and doesn't have the heart to rise again.

Epistrophe

The opposite of anaphora, with the repeated phrase placed at the end of each sentence. One of the best-known examples is in Abraham Lincoln's famous *Gettysburg Address* in which he speaks of government 'of the people, by the people, for the people'. Another appears in Paul's First Letter to the Corinthians in the King James Bible: 'When I was a child, I spake as a child, I understood as a child, I thought as a child.'

Symploce

When the repetition occurs at both the beginning and end of line. A much-quoted example is in Anne Morrow Lindbergh's book of 1955, *Gift from the Sea:* 'Perhaps this is the most important thing for me to take back from beach-living: simply the memory that each cycle of the tide is valid, each cycle of the wave is valid, each cycle of a relationship is valid.'

Essential tip

▓ Use groups of three for elegant emphasis.

The emotional journey

When you make a speech, your listeners place themselves in a heightened state of expectation. They follow you until you break the thread that connects you, but you do need to be aware of where you are taking them, in case the thread becomes tangled or breaks because you have strayed too far from your theme.

Read the short speech below and think about the succession of images as you are taken from one context to another. Think about where, on the social scale, your mind is taken. It's a speech to the undergraduates of a law college, on the occasion of the college's 175th anniversary.

My Lord Mayor, Mr Chairman, distinguished guests, ladies and gentlemen. It is a rare honour to be invited to address this celebration of your establishment's 175th anniversary. You have so much glory in your historic past. Your alumni, over the years, have gone on to great achievements, in government, in the law, and in literature. You have produced a Lord Chancellor, several Law Lords, two of them have been Prime Ministers, and one was even more distinguished as Leader of the Opposition.

I was travelling on a train to Newcastle, recently, when I got into conversation with another passenger, and I discovered that he, too, had qualified in this place, just as you are about to do. Something that he said struck a chord in me and I thought I'd share it with you.

He said he felt an obligation to make his legal knowledge available to those who most need it and can least afford it. The two often go hand in hand. Those who live in council estates with low levels of income and limited education often suffer injustice because they do not know where to find, and usually cannot afford, the best legal brains.

No doubt, in your studies, you will have come across such examples, and I would urge you to follow the example of my fellow train passenger and devote a part of your energies and commitment to pro bono *work.*

At the same time, I would also encourage you to stay close to your families, because as you rise to the top of your profession, as I am sure most of you will, there will be fierce pressures on you, and you will need the support and unconditional love that your families can provide.

This is a privileged place, an amazing building that houses the treasures of law, of knowledge and of art. Your library is one of the best of its kind in the land. And the walls of the principal rooms are graced with some exquisite works of art that the National Gallery would love to have. The main building itself is a work of art, a brilliant example of the best of early nineteenth century architecture. I'm sure it must imbue your bones with its elegance and history.

As you go into your new careers, eyes bright with hope, remember that your graduation does not mark the conclusion of your studies, merely the end of a chapter. There is much more to come, many more chapters to be read, and perhaps a few to be written as you make your mark in the world of law. I wish you all success.

Read it again, but this time draw a diagram to illustrate the mental journey. In the example, the opening is set quite high up the page, because it refers to dignitaries, but the second paragraph takes you to a train travelling to Newcastle. How does that feel, and was the change of scene too quick?

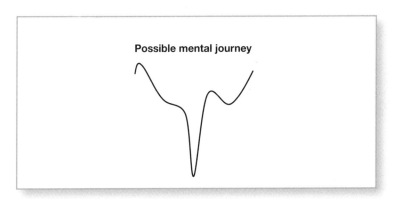

Possible mental journey

Consider how you feel when the speech refers to the poor and needy and *pro bono* work. What about the reference to family? Did it take you back to the days of living at home with your parents, perhaps with an untidy bedroom and tensions with siblings?

Whatever the effect on you, plot a diagram of where the speech takes you. Does it work, or did it cause you to lose track?

Remember this exercise, and plot similar diagrams of your own speeches or presentations in the future. You may know what you want to say, but you have to be consistent if you want your audience to follow you. Make it easy for them to understand and accept your Message, get excited about it, and be prepared to take some action as a result.

The next and final chapter is a *summary of the book*, to enable you to remind yourself of the essentials, and locate the right chapter for any of the points you want to re-read. It also repeats the **30-point checklist**, to help you identify where you might still have work to do, going forward.

For further reading I recommend:

Getting Your Point Across, Khan-Panni, How To Books, 2007.

Stand and Deliver: Leave Them Stirred not Shaken, Khan-Panni, Ecademy Press, 2010.

Summary

- The presentation's ten most important elements
- Six different kinds of presentation
- Language that creates images
- Oratorical devices that press the right buttons
- Mapping the emotional journey

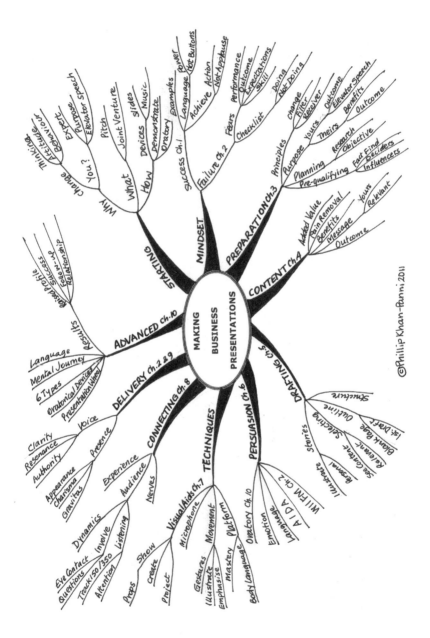

MAKING BUSINESS PRESENTATIONS

STARTING
Why — Change — Thinking — Attitude — Behaviour
You? — Expert — Purpose — Elevator speech
What — Pitch — Joint Venture
How — Devices — Slides — Music — Demonstrate — Oratory
— Examples

MINDSET
Success Ch.1 — Achieve — Language — Power — Hot Buttons
— Action
— Not Applause
Failure Ch.2 — Fears — Performance — Outcome
— Expectations
— Skill
— Checklist — Doing
— Not Doing

PREPARATION Ch.3
Principles — Change — Filler — Recover
Purpose — Yours — Outcome
— Theirs — Elevator Speech — Benefits — Outcome
Planning — Research — Objective
Pre-qualifying — Fact Find — Deciders — Influencers
— Relevant — Yours

CONTENT Ch.4
Added Value — Pain Removal
Benefits — Message — Yours
— Outcome — Relevant

DRAFTING Ch.5
Structure — Outline — 1st Draft — Blank Page
— Selecting — Reference — See Content
— Stories — Personal — Illustrate

PERSUASION Ch.6
WII FM Ch.2
A I D A
Language Ch.10 — Emotion

TECHNIQUES
Visual Aids Ch.7 — Microphone — Movement — Platform — Body Language
— Mastery
— Gestures — Illustrate — Emphasise
— Props — Create — Show — Project
— Oratory Ch.10

CONNECTING Ch.8
Audience — Nerves — Experience
— Dynamics — Eye Contact — Involve — Listening
— Questions — Track 50/350 — Attention

DELIVERY Ch.2 & 9
Voice — Clarity — Resonance — Authority
Presence — Appearance — Charisma — Gravitas

ADVANCED Ch.10
Results — Practice — Success — Follow-up — Relationship
Language
Mental Journey
6 Types
Oratorical Devices
Presentation Wheel

©Phillip Khan-Panni 2011

11

Going forward

In this chapter

- Summary of each chapter
- 30-point checklist to mark your progress
- How to get a free spoken summary of this book

Here's a **quick reminder** of the essential points made in each chapter. It will serve as a quick revision whenever you have a presentation to make, and will also help you to locate any topic that you wish to re-read.

The **30-point checklist** from Chapter 2 is also repeated here, to enable you to monitor your own progress. You may find it useful to run through it before each of the next two presentations you make, marking those elements that you feel certain are now in your own repertoire of skills, and leaving blank those that need more work.

Overcoming nervousness is always a major concern, and you'll find a seven-point plan to cope with that in Chapter 8.

Chapter summaries

Chapter 1 Successful presentations

- **Get emotional buy-in.** You have to get your audience *wanting* what you have to offer. Often a demonstration works well. It removes the risk of trial and can prove the benefit of your offering more strongly than a description can.

- Aim to **make change.** Plan to make some specific change in the thinking, attitude or behaviour of your audience. If there's to be no change, what's the point of the presentation?

- Try to **redefine your company's offering** in terms that relate to the business of the prospect. Do not describe what your offering *is*, but rather what it *does* . . . for the customer.

- Successful presentations **get action.** The best way to measure the effectiveness of a presentation is to see what people do as a result. If they do something new, your presentation has succeeded. So make it easy for them to take action, and tell them what they should do to benefit from your offering. When Aeschines spoke, they said, 'How well he spoke.' But when Demosthenes spoke, they said, 'Let's march!'

■ What '**demonstration**' can you use? If there is any way in which you can demonstrate your product or service in use, that would be a very powerful addition to your presentation.

Chapter 2 Why presentations fail

■ Check yourself against the **30-point checklist**. The checklist in Chapter 2 is to establish your present level against 30 essential components of successful presenting. The checklist will identify where you are at risk of falling short of your expectations. Later in this chapter the checklist is shown again, to enable you to mark your progress towards covering all the essentials.

■ Print out your answers and get help where indicated. After completing the checklist in Chapter 2, you will know where you might need help from a trainer or coach, who can then develop a programme specific to your needs.

■ Decide **what you are selling**. If you think about a demonstration of your offering, it could focus your mind more clearly on the real benefit, i.e. how it helps the customer.

■ Calculate the **cost of the presentation**. Do a rough calculation of the cost of researching, preparing and delivering the presentation, then add the cost of the time spent by all the people who attend. It's a sobering thought, isn't it? That's why you have to deliver good value, and expect a substantial return.

■ Refresh your memory of the **12 reasons why presentations fail**. This chapter lists the Dirty Dozen reasons for the failure of presentations. They fall within the three categories of content, style and delivery and include too much material, poor slides and being boring.

Chapter 3 Getting started

■ Develop an **Elevator Speech**. It will help your confidence as well as your focus if you are clear about your own role and what you bring to the table. The Elevator Speech is your 15-second statement of who you are and what you do, sufficient for someone to say, 'Tell me more.'

■ Decide on **the change you want** to bring about. Resist the temptation to start by listing all the attributes of your offering. Shift your attention to your audience and decide what you'd like to see changed as a result of your presentation. What action will you ask your audience to take to demonstrate that they have accepted your proposition?

▨ Follow the **4 Ps of customer-focused marketing**: Principles, Purpose, Planning and Pre-qualifying, and the rules for how to fact-find.

▨ **Answer the five questions for focus.** You need to have the answers to five questions on the tip of your tongue before you present: Why are you there? Why should the audience listen to you? What can you offer that they cannot get elsewhere? What do you want at the end? What's the least you will settle for?

▨ **Find out about your audience.** What are their concerns, their background, their status in the company, their role in the decision-making process, their expectations. One way or another, find out all you can about them beforehand. It's not safe to present blind. If your presentation is pitched at the wrong level, you could be wasting time.

Chapter 4 Deciding on what to say

▨ Start with the **PAT formula.** Purpose, Audience and Topic. You need to get all your ducks in a line, to ensure that your objective is consistent with the concerns of the people in your audience.

▨ **Different types of presentations.** Entertaining, Informative, Analytical, Problem Solving and Persuasive. In a sense, all of them need to be persuasive, because your basic requirement must be to get your listeners to accept what you are saying.

▨ **Know what you know.** People want to hear your take on the subject, not stuff they can get elsewhere. But they also want you to tell them something they already know, and how to use whatever new information or ideas you give them.

▨ **Decide on the problems** you can solve. List all the points you want to make and answer 'So what?' about each of those points.

▨ The **negative elements** of your presentation would be the Situation at present, its Weaknesses, and the Effect (**SWE**). The **positive elements** would be your Proposition, Reinforcement and the Action (**PRA**) you propose.

▨ **Project your brand.** What do you do for others through your product or service, what is the pain you remove, and do you keep the right company? Above all, know the one defining benefit of your brand – the one thing that identifies your business, and which makes people think of you when they encounter it.

Chapter 5 Drafting your presentation

■ **Following a structure.** It keeps you on track and makes it easy for your audience to follow you. There should be an overall structure, such as the approach, body and conclusion, and also a three-part structure for the development of your case in the body of the presentation.

■ **Structure helps impromptu speaking.** The same three-part structures (e.g. Past/Present/Future) would serve when you are asked to 'say a few words' or in Q&A.

■ **Headlines for focus.** Write a headline or two as though you were promoting your presentation as a public event. Notice the difference between a title and a headline. Is there a compelling benefit in the headline that will persuade people they need to attend?

■ **Blank page to first draft.** This is the technique that not only gets you out of trouble when you have an imminent deadline, it works as a first draft even when you have lots of time. It can enable you to arrive at a workable first draft in 15 to 20 minutes, if you know your topic.

■ **Talk about what you know.** Look within yourself for the message to impart. What is it that you really want others to hear from you and act upon? Can you solve a problem or remove a pain? Remember to promote your brand values at all times – the differences that distinguish you from your competitors and define your company.

Chapter 6 Being persuasive

■ The **seven essentials of persuasion**. Understand that persuasion is a process, and requires you to connect on the other person's level. The seven essentials are Listening, Relevance, Alternative, Meeting Expectations, Trust, AIDA and Commitment.

■ Check that your draft follows the **AIDA** sequence of Attention, Interest, Desire and Action. The graph on p. 85 explains the sequence and why you have to take the audience beyond the buying level.

■ **Standard patterns.** Even similar nations like the UK and USA have quite different expectations of the way a presentation should go, and different attention spans.

■ **Building trust.** Building on the four pillars: Reliability, Capability, Honesty and Empathy. In addition, Likeability.

Chapter 7 Visual aids

▨ **Which aids?** Decide on whether to use slides only or slides, flip chart and props, and arrange in advance to have them in position. Arrive early enough to check that they are available and working.

▨ **Slides.** Minimise content of all slides, and the number of slides. Work to no more than one per minute. Use sans serif fonts, in large sizes, and follow the 5×5 rule (lines/words per line), and pictures whenever possible. Have a numbered printout of slides, blank the screen with the letter B. Minimise transitions, fonts and gimmicks.

▨ **Video.** If using film or video, embed it into a slide and practise using it. Keep it short and professionally produced, topped and tailed. Convert all video clips to WMV format and have a sound-track.

▨ **Flip chart.** If using a flip chart, write or draw in pencil in advance. Always use letters at least 5cm (2in) in height, and write with broad nib (chisel tip) pens. Do not use if there are more than 30 people present. The same applies to white boards.

▨ **Rehearse, rehearse, rehearse.**

Chapter 8 Connecting with the audience

▨ **Nervousness.** Overcoming the fear of public speaking is top of most people's agendas. Here you will find a seven-point strategy to cope with performance anxiety.

▨ **Listening.** Audiences trend to drift in and out of paying attention. To help them catch up, add looping back transitions to your draft, to remind them of where you have been and where you are going next.

▨ **Test** your draft on someone not familiar with your subject, to see if you are easy to understand. Remember that different cultures have different reasoning styles, but all audiences could contain some who are hostile, some passive, and others who want value.

▨ **Association and conditioning.** Take account of the way people have been conditioned to listen or watch, and use colour and images to aid understanding and recall.

▨ **Group dynamics.** When presenting to a group, be aware of how audiences process information, and indulge in collective behaviour, either for or against you.

▨ **Facts.** Pass all your facts through your personal filter to make them your Wisdom. Ask yourself, are you delivering the words or the message?

▨ **Record** two three-minute talks on camera, one impromptu, one scripted. Listen for the level of conviction in your voice.

Chapter 9 Delivering your presentation

- **Power in delivery.** Avoid building up the expectation of a confrontation. Focus, instead, on how the listeners do business, and offer to support that. It makes you a powerful ally rather than an opponent.

- **Appearance and charisma.** Your 'stage presence' depends on the way you look, your comportment and your personal authority. You will be judged within the first ten seconds of rising to speak.

- **Trust.** Make a commitment to helping the audience solve their problems and do better business. You are expecting them to trust you on a single hearing. Focus on them.

- **Voice.** The way you sound can make or break a presentation. Think of your own reaction to broadcasters, for example. You *can* change the way you sound and improve your resonance. There's a simple exercise that you can do.

- **Microphone.** Nine essential microphone techniques and using them to reach every member of the audience.

- **Platform skills.** Take charge of your space, and move with a purpose. Choose the right pitch, vary your pace, use pauses. Practise with a voice recorder.

- **Pitch, pace and pauses.** How to add variety to the way you speak from the platform.

- **Gestures and movement.** Use expressive gestures that enhance your meaning, and be aware of how you are being perceived. Body language should be confident and approachable.

- **Presentation essentials.** A 15-point summary of all you need to know about delivering a presentation that will make an impact and get results.

Chapter 10 Advanced

- **Presentation wheel.** On a photocopy of the wheel, mark the scores you would give these ten elements in a presentation, and see how they differ from a colleague's scores. Check again after several presentations, to see if your opinion has changed.

- **HAMO IS DEAF**, which stands for Hook, Audience, 3Ms, Objective, Interest Building, Structure, Delivery, Emotion, Action Call, Facts and Figures. List them in order of importance, and see if your opinion changes after a few presentations.

- **Advanced types of presentations.** These have different objectives and follow different rules, but all need to be persuasive. They are New business, Persuasive, Motivational, Entertaining, Ceremonial and Instructional.

- **Language.** The language you use will determine how well your message is understood. Abstract terms are hard to visualise, and there is no picture for 'Don't'.

- **Oratory.** How top speakers use linguistic devices and parts of speech to dramatise, build to a climax and stimulate emotional responses. Skilful orators like Obama use them.

- **Emotional journey.** Draw a diagram to represent the imaginary journey of your presentation. Is it a congruent one, or will you lose your listeners along the way? When you have mastered this process, above all, you will be an advanced presenter.

Checking your progress

Before the next two presentations you make, run through this checklist again and see how many of the items you can tick. Leave blank those that you have yet to master. They will remind you of the work you have still to do in developing your presentation skills.

30-point checklist

Why should people listen to you?

---

1   Do you have a 15-second Elevator Speech?

2   Are you an acknowledged expert in your subject?

3   Do you always speak with passion and conviction from the platform?

4   Is your speech or presentation always focused on making some change?

5   Can you stand and 'say a few words' at a moment's notice?

What makes you a 'must have'?

---

6   Can you state your USP in ten words or less?

7   Can you list five things about yourself or your business that distinguish you from competitors?

8   Would you pay to hear you speak?

9   Do you always get good, positive feedback when you speak or present?

10 Do you make your proposition in a compelling, dramatic way?

Preparation

11 Do you follow a well-defined structure that helps listeners stay on track with you?

12 Can you go from blank page to first draft in 15 minutes?

13 Have you recently had coaching or training in speaking or presentation skills?

14 Do you open with a memorable hook?

15 Can you easily speak the sentences you write in your script?

Using visual aids

16 Can you present without PowerPoint?

17 Do your slides carry a maximum of 30 words?

18 Do you know how to blank the screen with a single key?

19 Can you skip to any slide without scrolling?

20 Do you present without looking at the screen behind you?

Being persuasive

21 Do you follow the AIDA sequence of persuasion?

22 Do you identify and address the needs of your audience?

23 Do you usually get business as a result of your presentations?

24 Do you tell your audience what to do at the end of your speech or presentation?

25 Do you usually get questions when you have finished speaking?

Connecting with the audience

26 Do you like the way you look when presenting?

27 Do you like the way you sound when presenting?

28 Do you ask rhetorical questions during your speech or presentation?

29 Do you use oratorical devices such as triads and anaphora?

30 Do you usually get a good connection with the audience, with hearty applause?

Spoken summary

You can download a free ten-track spoken summary of this book, and burn it onto a CD to play in your car or elsewhere. Go to **www. pkpcommunicators.com/presentationbook**, and use the code 250956 as the password.

Index

FINANCIAL TIMES
··
Essential Guides

Also in this series

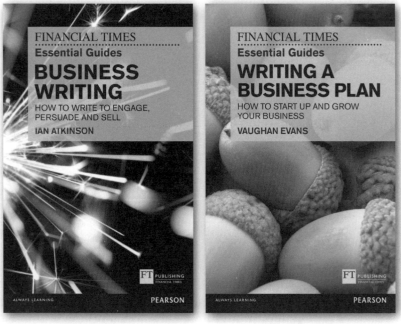

FINANCIAL TIMES
Essential Guides
**BUSINESS
WRITING**
HOW TO WRITE TO ENGAGE,
PERSUADE AND SELL
IAN ATKINSON

FINANCIAL TIMES
Essential Guides
**WRITING A
BUSINESS PLAN**
HOW TO START UP AND GROW
YOUR BUSINESS
VAUGHAN EVANS

FT PUBLISHING
FINANCIAL TIMES

ALWAYS LEARNING   PEARSON

FT PUBLISHING
FINANCIAL TIMES

ALWAYS LEARNING   PEARSON

9780273761136   9780273757986

Available to buy online and from all good bookshops
www.pearson-books.com